CONTENTS

The compiler of this book wishes to express his most sincere thanks to 'ELECTRONICS AUSTRALIA' the leading radio, T.V. and electronics magazine published in that continent, and to Syndication International Ltd., their agents, for permission to use much of the material in this book which appeared originally as articles in that magazine.

B. B. BABANI
LONDON - JUNE 1973

CONSTRUCTORS MANUAL OF ELECTRONIC CIRCUITS FOR THE HOME

BY

B.B.BABANI

BABANI PRESS
The Publishing Division of
Babani Trading and Finance Co. Limited
The Grampians,
Shepherds Bush Road,
London W6 7NF

Although every care is taken with the preparation of this book the publishers will not be responsible for any errors that might occur

I.S.B.N. 0 85934 015 5

Printed in Belgium

HOT SEAT

OCCASIONALLY STAFF LET THEIR HAIR DOWN AND BECOME INVOLVED IN SOMETHING DESIGNED PURELY FOR LAUGHS. IT IS IN THIS MOOD THAT WE PRESENT THE "HOT SEAT", A PARTY GADGET WHICH SHOULD HAVE A WIDE APPEAL, PARTICULARLY AMOUNG THE YOUNG FOLK AND THOSE ASSOCIATED WITH YOUTH ACTIVITIES.

Party gadgets and "crowd stoppers" have always been popular, particularly where an element of competition exists. The device described presents an air of challenge, even to those who fail miserably in their attempts to beat or outwit it.

The Hot Seat is a simple reaction time comparison device whereby two contestants can match their reaction times one against the other. There is nothing really new about this. Television and radio quiz programs have long used electronic devices to establish, without room for question, the first contestant to respond to a particular stimulus. The stimulus, usually, is the possing of a question, and the reaction time is that needed for the contestant to decide that he can answer it.

Alternatively, a simple visual stimulus may be used, such as the turning on of a light or the changing of a light from red to green. In this case, the reaction time is the purely physical one which is a natural limitation of all human nervous systems. This is a measure of alertness, amoung other things, and can win or lose many simple games or contests.

The novel feature of the Hot Seat is the manner in which the loser is encouraged to announce this fact. The two contestants are seated on special chairs or stools, and face a simple control panel on which are a pair of lamps, one red and one green. Each contestant is provided with a press button and told to press his button immediately the light changes from red to green. The contestant who is fortunate enough to press his button first is allowed to remain seated; the loser is propelled to his feet with a well delivered electric shock in the seat of the pants!

We can vouch for this fact that this gadget will make any party go with a bang.

Where any degree of electric shock is involved, many people are unduly fearful, and may take a lot of convincing that such a shock can be, at one and the same time, uncomfortable yet perfectly safe.
The Hot Seat delivers its message in the form of a short sharp pulse. While leaving no doubt in the mind of the victim that he has been bitten, it does so in a manner which is perfectly harmless.

In considering the safety of such a device, four major factors need to be considered. These are:

(1) The voltage of the pulse
(2) The energy content of the pulse
(3) The duration of the pulse
(4) The repetition rate of the pulse

As a guide to the design of our unit we took the safety specifications for electric fences. These are particularly stringent and conservative, as well as being based on "worst case" conditions far in excess of any likely to be encountered in an application such as ours.

The figures laid down, together with those for our own device, are as follows:

PULSE	FENCE	HOT SEAT
Length	100mS	3uS
Energy	2.5mC	0.15mC
Voltage	5KV*	4.5KV*
Time between pulses	0.75S	1S

* Across 1M load

With regard to the last item, the figure of 1 second for the Hot Seat is the shortest time between pulses which could occur but, in practice, the device delivers only one pulse per contest. There is no automatic repetition rate.

In addition we must consider the manner in which a shock might be applied to the body, either deliberately or accidentally. The deliberate application is to the thighs and this in itself provides a high safety factor. Thus applied it does not direct a flow of current across the heart area, nor does it permit the person to grasp the contacts as the result of any muscular contraction, even assuming that these latter were energised for long enough to constitute a hazard.

Finally, to remove the last element of risk, we operate the Hot Seat from batteries, thus eliminating any possibility that the mains might become involved, no matter how remote this possibility might be.

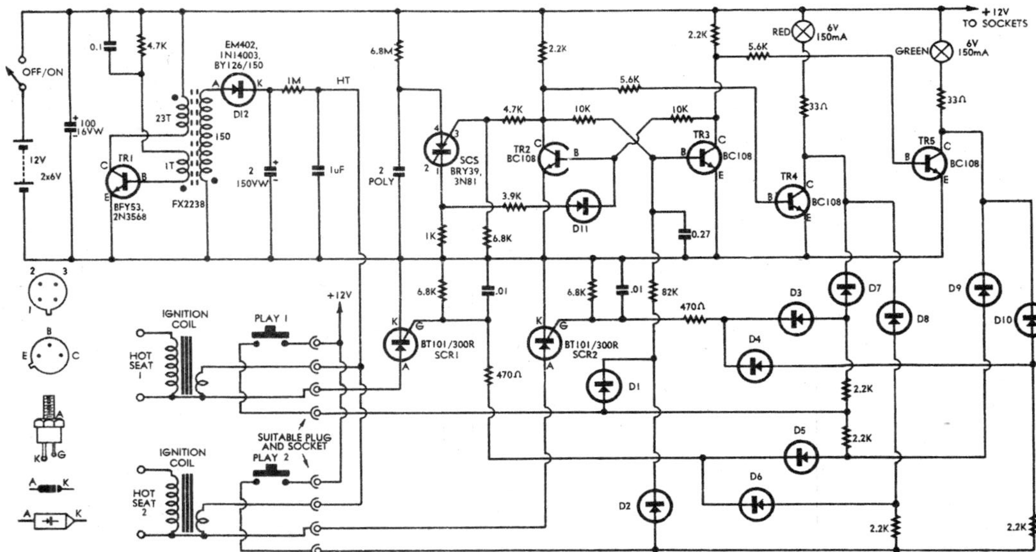

HOT SEAT

The operation of each circuit section is explained in the article. Study the circuit carefully in conjunction with the text.

ALL SMALL SIGNAL DIODES
BA100, OA202, OR SIMILAR

So that the time of pulse, voltage and energy level could be predicted, we decided to use the now well-known automotive CD ignition circuit. With this circuit it is easy to scale down the voltage and energy to safe levels.

Following the dump capacitor, the circuit divides into two identical sections involving two thyristors, two ignition coils, and two interconnected triggering circuits.

These two circuits share the one dump capacitor, and it is obvious that the discharge can be applied to only one coil at a time, depending on which thyristor is triggered first.

Apart from the DC power supply, we can divide the circuit into four parts : A timer, logic flip-flop, trigger circuit 1, trigger circuit 2.

The timer, set for a countldown of approximately 30 seconds, uses a BRY39 SCS (silicon controlled switch) as its active device.

When activated, the 2uF polyester capacitor is slowly charged via the 6.8M resistor towards the 12 volt supply rail. Because the SCS anode gate is connected to a voltage divider which sets its potential at about half voltage (6 volts), the anode (connected to the junction of the resistor and capacitor) can reach a potential only about 600mV higher than this, at which point the device "fires".

This discharges the capacitor via the SCS, producing a pulse of about 5V (positive at the cathode) across the 1K cathode resistor. This pulse is fed to the logic flip-flop to change its state.

The logic flip-flop is a bistable multivibrator circuit formed around a pair of BC108 (or similar) transistors (TR2 and TR3). Its state is controlled by pulses from the timer or chair button switching circuits.

The positive pulse from the SCS cathode is fed to the base of TR2 which, previously being cut off, is now switched to saturation. This causes the collector voltage of TR2 to fall towards the negative rail (to typically, about IV positive) bringing the anode gate of the SCS down with it. With both the anode and anode gate at a low voltage, the SCS is held off, preparing the way for another charging cycle.

The collectors of TR2 and TR3 are connected to the bases of TR4 and TR5 respectively. These two latter transistors control the red and green lamps used to provide the visual stimulus. They also connect to the diode matrix which forms part of the decision-making circuitry.

It is more satisfactory to drive the diode matrix from TR4 and TR5, rather than directly from the flip-flop. We can achieve lower circuit clamping resistance and lower collector to emitter potential (higher collector saturation current) by using TR4 and TR5 than we could from TR2 and TR3, even though they are the same transistor types.

The lamps are connected as collector loads for TR4 and TR5; red in TR4 and green in TR5. The 33 ohm resistors in series with the lamps are to limit the collector current at the moment of switching on, when the resistance of a cold lamp filament is low.

If the flip-flop was in a state where TR2 was cut off and TR3 saturated, the high and low potentials of their respective collectors would cause TR4 to be saturated (red light on) and TR5 to be cut off (green light off).

When the timer capacitor discharges and switches TR2 into saturation, TR3 is cut off. TR2 now cuts off TR4 (red light off) and TR3 switches TR5 into saturation (green light on).

As already stated, the flip-flop is interconnected with the decision-making circuitry. The reason for this is to provide an "anti-cheat" function, against the possibility that a player will try to beat his opponent by pressing his button before the light changes from red to green.

To forestall this, the circuit is so arranged that, while the flip-flop is in the "red" mode, the trigger circuits are transposed. Thus, any competitor who presses his button before the lights change, will direct the shock to his own posterior, rather than to that of his opponent. Needless to say, not many players try to cheat once they appreciate the implications.

The diode matrix may be divided into two trigger circuits; trigger circuit 1 and trigger circuit 2. Trigger circuit 1 is controlled by the button "Play 1" and consists of diodes D3, D7, D5 and D9. Trigger circuit 2, controlled by button "Play 2" consists of diodes D4, D10, D6 and D8.

When the flip-flop is in the "red" state trigger circuit 1 is capable of triggering SCR1; when in the "green" state, trigger circuit 1 is capable of triggering SCR2. The reverse applies to trigger circuit 2.

Referring to the circuit diagram, if we press "Play 1" button, 12 volts is applied to a pair of 2.2K resistors connected to diodes D3, D7, D5, D9 and directly to D1. Because TR4 is saturated and its collector potential is low (about 1 volt), D7 will be forward biased and will remove any trigger potential capable of triggering SCR2 via D3.

At the same time, TR5's high collector potential will reverse bias diode D9, allowing current to flow via D5 into SCR1's gate circuit causing it to trigger.

The same reasoning could be applied to SCR2's trigger circuit via D4 from "Play 2" button if it were depressed.

The purpose of D1 and D2 is to re-set the flip-flop into the "red" mode after each play. As mentioned earlier, D1 is connected directly to the 12 volt source via the "Play 1" button when this is depressed. This diode

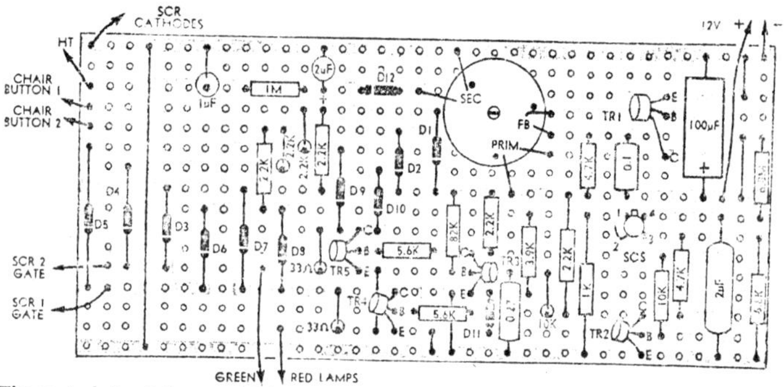

The control circuit is assembled on a piece of veroboard, cut to the pattern shown on the grey overprint. The location of the components is shown in relation to this pattern, being viewed from the components side.

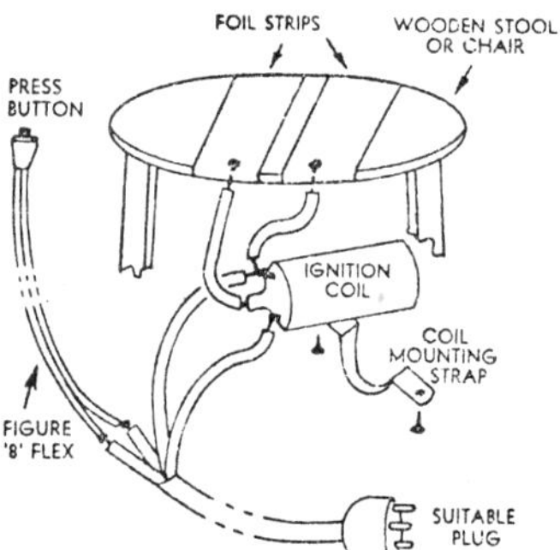

Details of the stool contacts and wiring associated with them. The ignition coil is mounted under the seat.

is then forward biased, and introduces a positive going pulse, limited by the series 82K resistor, to the base circuit of TR3.

When the green lamp is illuminated (i.e. after the countdown), TR3's base circuit is "low", and will thus accept a positive-going trigger pulse from either "Play 1" or "Play 2" via the diodes D1 and D2 and the 82K resistor. When the positive-going pulse arrives, the state of the flip-flop will be reversed - TR2 cut off and TR3 saturate.

As well as switching the lamps back to the "red" mode, the fact that TR2 is cut off means that its collector moves towards the positive rail potential (12V), taking the SCS anode gate up proportionally with it. Thus the SCS anode gate up proportionally with it. Thus the SCS is restored to its original condition, whereby the timing cycle can re-commence.

In the event that the system is unattended when the timer switches from red to green, and neither play button is pressed, the system will remain in the green mode indefinitely. To reset it for another pair of contestants it will be necessary to press one of the play buttons, being careful not to deliver an unwarranted shock to any contestant who may be already seated.

At this point, it should be fairly obvious as to how the game is played.

We have two opponents "A" and "B" seated on their respective chairs or stools. If we assume that the timer is timing out (the RED indicator illuminated), the trigger circuit of "A's" chair will be operable by "A's" button (of "B" with B's" button). If, during this period, "A" or "B" depresses his button, the player responsible will receive the charge - with the appropriate surprised reaction!

If either contestant does not trigger the system before the timer times out, the flip-flop, through the diode network, will arm "A's" trigger button with "B's" trigger circuit, and vice versa. Whoever triggers the system first will have the delight of seeing his opponent stand to attention in no uncertain manner!

Although we have nominated currently available transistor types, there is no reason why some experiment cannot be done with disposal "computer" types readily obtainable for a moderate cost from various sources. NPN silicon types are preferable, but NPN germanium types would probably work just as well. Suitable diodes could also come from the same source.

The main disadvantage of using computer transistors is that more often than not, the leads on these devices are extremely short, and extension wires will have to be soldered to them to make comfortable connection to the Veroboard circuit board. Similar treatment may have to be given to the small diodes extracted from similar boards.

Each seat, in our case, was made from a robust wooden stool. The legs were shortened so that the normal seating position was optimum for a fast stand. Under each of these a 12 volt automotive ignition coil was mounted by a simple metal strap made for crating strip. The secondary of each coil is wired to two aluminium foil strips placed about six inches apart in the sitting area. Connection to the foil is made by flat headed nails driven through the stool from the top.

We found that the best source of suitable ignition coils was a motor wrecking yard.

There is little advantage to be gained in using new ignition coils, as the money spent on a pair of new coils will possibly buy all the components, including the two second-hand coils.

Although the layout drawing of one "hot seat" shows a retaining strap wrapped around the body of the ignition coil, you may or may not be able to obtain the coil's original mounting clip with the coil. One of our two coils was equipped with a clip, and one without. The wrecking yard may include a clip on each coil only if you ask for it, as most dealers assume that such coils are bought as stand-ins for motor vehicles, and that the coils would be mounted in the clip already in the vehicle.

A cable runs from each stool to the central control box. The cable can be made from two lengths of "figure eight" flex, preferably in distinctive colours for each of identification. The two lengths may be sheathed in 4mm PVC tubing for additional protection.

One pair connects to the primary circuit of the ignition coil, the other pair to a short flexible lead terminating in the player's trigger button.

The trigger buttons are simple affairs similar to the types used beside hospital beds or attached to slide projector control cables. A refrigerator door switch (normally open type) makes an excellent button, and may be housed in a discarded plastic pill bottle.

Indicator lamps used in the prototype were 150mA, 6V ones with coloured "stop" and "go" windows placed in front of them. You may be fortunate to "score" a couple of computer indicators. These may be equipped with 24 volt or 50 volt lamps, which would have to be replaced.

The last details to attend to is the type of housing for the electronics and battery. Any large rectangular biscuit or chocolate tin would be suitable, or a large cigar box. Even disposal equipment cases may prove useful for this role.

Whatever is used, the layout is not critical.

PARTS LIST

2 Wooden stools or chairs (see text)
2 12 volt ignition coils
2 6 volt lantern batteries
2 Miniature 4 pin speaker sockets with matching plugs
2 Pushbutton switches (Normally open)
1 Single pole toggle switch
2 Indicator lamps, one red, one green (see text)
1 Carrying case with handle
2 FX2238 pot core halves
2oz 34 s.w.g. enamel copper wire
1oz 28 s.w.g. enamel copper wire
2 BT101-300R or similar thyristors
2BFY53, 2N3568 or similar
4 BC108, BC208 or similar
1 BRY39 or 3N81
11 BA100, OA202 or similar small signal diode
1 BY126-150 or similar silicon rectifier diode

RESISTORS ($\frac{1}{2}$ watt, unless specified)

1 6.8M	2 4.7K
1 1M	1 3.9K
1 82K	6 2.2K
2 10K	1 1K
3 6.8K	2 470 ohm
2 5.6K	2 33ohm

CAPACITORS (Polyester 160V unless specified)

1 100uF 16VW electrolytic	1 0.27uF
1 2uF 150VW electrolytic	1 0.1uF
1 2uF (or 2.2uF)	2 0.01uF
1 1uF	

1 Piece Veroboard (0.15") $5\frac{1}{2}$" x $2\frac{1}{2}$"
8 yds brown figure 8 power flex
12 yards white figure 8 power flex
8 yds 4mm plastic sleeving
Hookup wire, nuts, bolts, packing strap, flat head nails aluminium foil strip etc.

NOTE : Resistor wattage ratings and capacitor voltage ratings are those used for our prototype. Components with higher ratings may generally be used providing they are physically compatible. Components with low ratings may also be used in some cases, provided the ratings are not exceeded.

CONVERTER TRANSFORMER SPECIFICATIONS:

Secondary: (Wound on first) 160 turns 34 s.w.g. enamel wire. Code start with one knot. Fit 1mmPVC sleeving over lead-outs.
Interleave: One wrap polythene film cut slightly wider than inside width of bobbin.
Primary: 23 turns 28 s.w.g. enamel wire. Code start with two knots. Place 1mm sleeving over lead-outs.
Feedback: 1 turn any gauge. Code start with three knots.
Outer wrap: One wrap of electrical tape.
Cores: Two FX2238 pot core halves.

ADAPT YOUR RADIO TO DOUBLE AS AN INTERCOM

DETAILS OF SIMPLE ADAPTATION CIRCUITRY WHICH WILL ALLOW ALMOST ANY DOMESTIC RECEIVER TO DOUBLE AS TWO-WAY INTERCOM OR MONITOR SYSTEM. JUST THE IDEA FOR PROVIDING AN AUDIO LINK BETWEEN THE HOUSE AND A GARAGE OR "HAM SHACK"

If you want to provide a two-way audio link between your house and garage or workshop, or to give your wife a monitor system which will allow her to listen for baby's wail while she is in the kitchen, there's probably no need for you to buy or build a complete intercom set for the job. Most likely you already possess about half of the necessary hardware of such a system, in potential form: the audio amplifier section and the loud speaker of a standard radio receiver.

All one has to do, in order to allow the radio to double as a useful intercom or monitor, is add the simple adaptor circuitry described in this article. The conversion is straight forward, requires little effort, and - most importantly - will involve a minimum outlay. With some receivers, the cost will be no more than £2.50 or even less if some of the parts are salvaged from the "junk box".

The adaptor circuitry consists basically of a remote speaker, a step-up audio transformer, and a simple switching system incorporating a relay. In addition to these may be required a small one-transistor preamp, in cases where the receiver audio section has inadequate gain, together with a small auxiliary power supply necessary if the relay cannot be energised by the receiver power supply.

The full schematic of the adaptor circuitry is shown in the main diagram, with the basic sections of the existing receiver shown in block form. As may be seen, there are only two points at which are connections to the existing circuitry, one being at the rotor of the volume control pot while the other is at the loudspeaker voice coil.

There are two switch elements connected to the receiver circuitry at the volume control rotor and audio section input; one element is part of the "local" intercom function switch, while the other is a changeover contact set actuated by the relay - which performs the "remote" function switching. Both switch elements are connected such that when both "local" and "remote" stations are in the "listen" mode, the tuner output is fed to the audio section in the normal fashion. However, if either station is switched to the "talk" mode, or the "local" station is switched to the "monitor" mode, the tuner output is disconnected and signals from the appropriate station connected to the audio section input.

At the receiver voice coil circuit there are three switch elements, two of which form the remainder of the "local" intercome function switch while the third is a second changeover contact set actuated by the "remote" relay. The first and third switch elements are used for switching of the actual speakers; when both stations are in the "listen" mode, the speakers are connected in parallel across the receiver audio output. In this mode the system thus behaves purely as a radio receiver, the "remote" intercom station acting as an extension speaker.

When either station is switched to the "talk" mode, the corresponding speaker is disconnected from the output of the audio section and routed to the input, via a reverse-connected speaker transformer, the preamp (if used) and the input switch elements. Hence, either station can communicate at will to the other, by interrupting the radio program material.

A separate gain control fitted to the intercom input circuitry allows the receiver volume control to be adjusted quite independently of the intercom system, so that if desired the radio program may be turned well down or even right off without affecting the intercom volume or sensitivity.

The "local" function switch is a 3-pole three-position switch, preferably one in which one position ("talk") is provided with a spring-return action.

The third position of the "local" switch is used to provide the "remote monitor" function. By causing the remote station speaker to be connected to the input of the system, this position allows the unit to be used for monitoring and "baby minding" under full control of the local station. Hence, mother can enjoy her favourite musical program or soap-opera in the kitchen, interrupting it only briefly every now and again to listen for possible "trouble".

The relay, which performs "remote" station-switching is a low voltage, low current type, operated by a simple "press to talk" push-button at the remote station itself or alternatively by the third pole of the "local" switch in the "monitor" position. As may be seen, the use of a relay simplifies the intercom-switching and allows the use of relatively economical two-core shielded cable to the remote station.

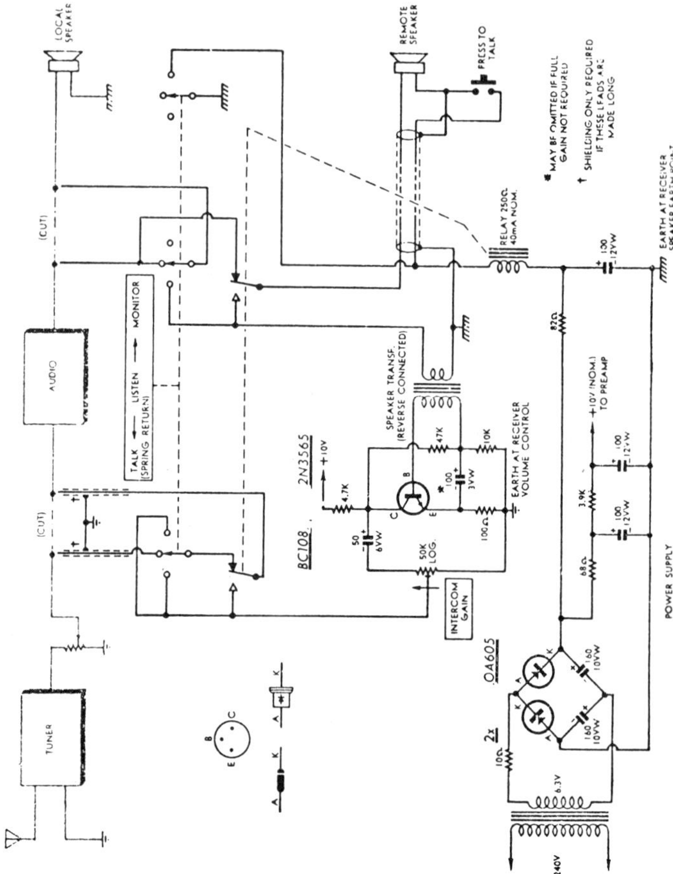
LOCAL SPEAKER
REMOTE SPEAKER
PRESS TO TALK
* MAY BE OMITTED IF FULL GAIN NOT REQUIRED
† SHIELDING ONLY REQUIRED IF THESE LEADS ARE MADE LONG
RELAY 250Ω 40mA NOM.
100 12VW
EARTH AT RECEIVER SPEAKER EARTH POINT
(CUT)
(CUT)
TALK ← LISTEN → MONITOR
(SPRING RETURN)
AUDIO
TUNER
SPEAKER TRANSF. (REVERSE CONNECTED)
47K
10K
100 3VW
EARTH AT RECEIVER VOLUME CONTROL
100Ω
2N3565
+10V
4.7K
B
C
E
BC108
50 6VW
50K LOG.
INTERCOM GAIN
82Ω
+10V (NOM.) TO PREAMP
100 12VW
3.9K
100 12VW
68Ω
POWER SUPPLY (IF REQUIRED)
OA605
2x
160 10VW
160 10VW
K
A
10Ω
6.3V
240V

Although almost any relay providing two changeover contact sets could be used, units having either very high or very low resistance coils would pose power supply problems; for this reason we are suggesting a relay having a nominal 250 ohm coil and requiring a nominal actuating current of around 40mA.

The step-up transformer used to couple the intercom speakers to the input of the receiver audio amplifier may conveniently be a standard speaker transformer used back-to-front. The impedance ratio of the transformer is not especially critical, although for best efficiency it should have a fairly high ratio. A small low-powered rating transformer willbe quite adequate, and will in fact be an advantage as the transformer must be kept away from the receiver power and output transformers to obviate hum and magnetic feedback problems. If close spacing is unavoidable, the transformer may have to be orientated for minimum coupling.

In cases where the gain of the receiver audio section is sufficiently high, the secondary of the step-up transformer may be connected directly to the intercom gain control. However, with receivers having a single audio stage, this connection will most likely give inadequate intercom gain, and it will be necessary to employ the simple, one-transistor preamp stage shown. The transistor employed may be almost any general-purpose NPN silicon type, suitable devices being the BC108.

The pre-amp circuit shown is a simple common-emitter stage using transformer input coupling. If the full gain is not required, the 100uF 3VW electrolytic between the emitter and the "cold" end of the transformer secondary may be omitted. Power requirements for the preampl are a modest 10V (approx.) at 1mA, which may be derived either from the receiver circuitry or from the relay power supply shown.

The power supply itself is a simple "full-wave" voltage doubler circuit, connected to a source of 6.3V AC. The AC may be taken from a heater winding on the receiver transformer, if this is convenient, or alternatively from a separate transformer as shown. If a separate transformer is used, it should preferably be a small or miniature type (the current drain will be no more than 45 - 50mA at most).

Where the local and remote stations are to be separated by a long distance, it is recommended that the remote speaker be of 15 ohms impedance to reduce cable resistance losses. In such cases it may also be advantageous to fit a 15 ohm speaker to the receiver, to balance the gain of the intercom in both directions. Although this may introduce some mismatch into the receiver output coupling the effect upon performance will in most cases be negligible.

MET SYSTEM

THIS DESCRIPTION IS OF A METEOROLOGICAL CONSOLE FROM WHICH WE CAN READ OFF WIND DIRECTION, WIND SPEED, AND WET AND DRY BULB SHADE MEASUREMENTS FOR TEMPERATURE AND HUMIDITY.

The wind direction indicator consists of a pair of selsyn transmitters currently available from surplus dealers. Run them on 22V for better stability of the slave. Sensitivity will not suffer.

The wind vane is cut from a piece of galvanised sheet and slotted into a piece of 3/8in copper tubing and silver soldered in place. The point of balance was ascertained and the vane silver soldered to the shaft of the selsyn. A 2 ounce tobacco tin was mounted between the extension and the shaft to weather protect the bearing. Before installation, the whole thing was given several coats of good rust preventing paint.

To mount the slave, cut a 2½in diameter hole through a 1½in thick block of pine with a hole saw, to pass the shaft and the extension. Drill two holes through opposite sides of the front cast-iron mounting ring. The selsyn is screwed to the wood block with 2in wood screws. The small amount of shaft extension protruding is then sawn off, and the wood block, complete with selsyn, is screwed to the front panel of the console, making sure that the hole for the pointer is accurately located.

The pointer was made from a piece of galvanised sheet, and silver soldered to a shaft of the same diameter as the inside of the extension.

For the wind speed indicator use only 50 mph scale, as this is all that is needed.

Those interested in cutting costs may care to know that any of the old-style glass-cased transistors with the paint scraped off will work satisfactorily in place of the OCP71, in this instance. A light dependent resistor such as the ORP12, LDR03 or the RPY 25 will also perform well.

The active elements of the thermometers are two Post Office type 1AN thermistors, one for the dry bulb reading and one for the wet bulb reading. The diagram shows how these are set up. Wet bulb readings are taken by pressing a button. The meter is a 50uA model about 3in square. Calibration is simple: if 50uA is taken as 100 degrees F, then the scale is linear down to at least 40 degrees F. This works out at two degrees per microamp. We added five more divisions to the top end of the scale on our meter to increase readings.

Circuits of the three sections, showing the arrangement of each and their connection to a common power supply.

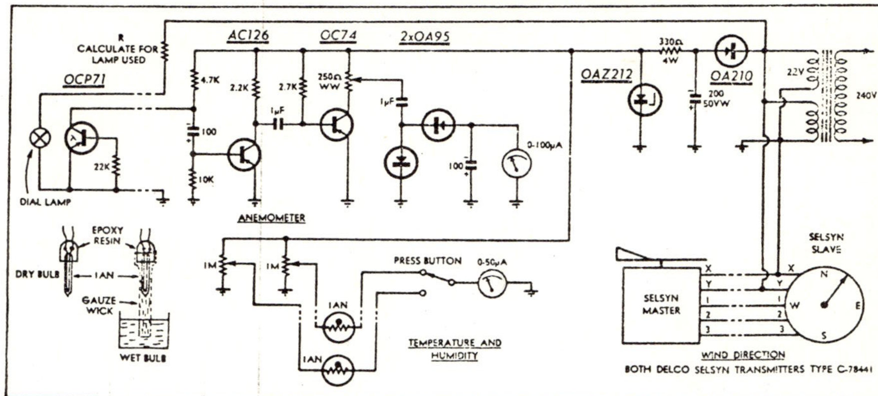

To calibrate, place both 1ANs in a mixing bowl with a thermometer of known accuracy, pour in water heated to just over 100 degrees F, wait for the temperature to drop to exactly 100 degrees and set the 1M calibrating potentiometer until the temperature indicated by both thermistors is 100 degrees. If the results of other readers are the same as mine, indications will then be perfectly accurate down to 40 degrees. This can and should be checked by replacing the warm water with water at 40 degrees. A check can be made at this temperature, and then all the way up the scale by gradually adding hot water.

A COMBINATION LOCK

The combination lock has been around for a long time. Most of us would have been movies of "expert" safecrackers with their ears to the safe, listening as the tumblers clicked into place and then dramatically opening the door to reveal the inner treasures.

Maybe the tumblers are not all that easy to hear in reality, a fact which movie makers would doubtless ignore. Nevertheless, a lock which is completely silent in operation, and which can be built from readily available electronic components, makes an interesting project.

We plan to present a lock such as this. As well as the features mentioned above, it has the added advantage of an alarm function which would sound at the first wrong move. Also, the combination is relatively easy to change - unlike the mechanical variety.

While presented basically as a beginner's exercise, the project could have more serious applications. Suitably engineered it could form the basis of a genuine security lock, and we will have more to say about the actual lock mechanism later on.

At a less serious level it could form the basis of a game of chance at a typical charity fair. Contestants could be offered so many tries for, say, 10p and anyone fortunate enough to "crack" the combination and open the safe would receive the prize it contained. Considering the odds, the organisers could afford to be fairly generous in regard to the prize.

Let us look at the broad concepts of such a lock. Basically, we need a device which will allow a mechnical device to activate when, and only when, the information given it is correct in every respect. If any of the information is incorrect, the device should be able to warn of this fact.

A very simple system of logic circuits is able to do this. If the term logic is new to our younger readers, it is the field which embraces number systems and control using electronics. Naturally, computers fall into this class. In fact, some of the things in the electronic combination lock will be found in every computer.

COMBINATION LOCK

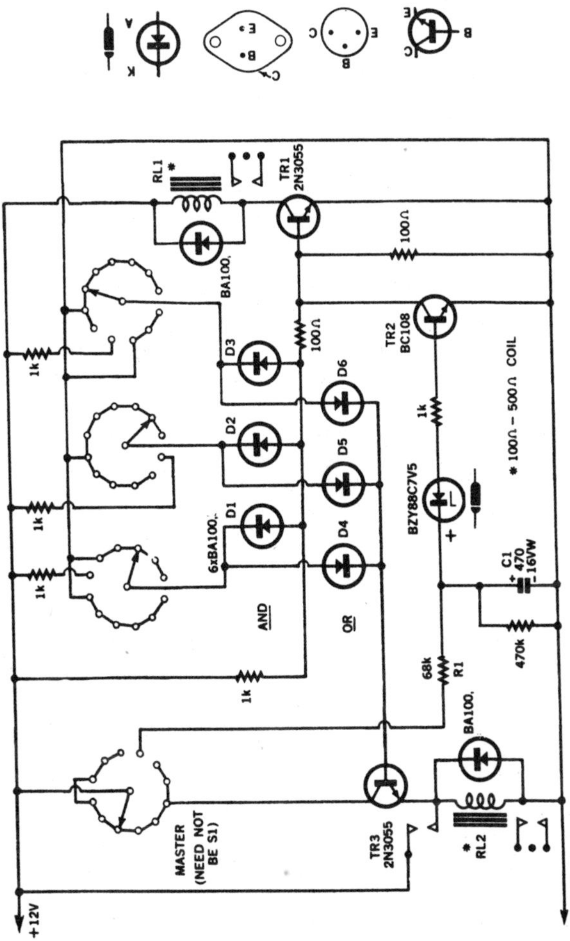

Circuit of the electronic combination system. If an actual lock is not required, the supply voltage may be increased somewhat to allow higher resistance relays to be used.

These are the so-called "gates". As the name implies, a gate is a two state device - it can be only open or shut. The state of the gate depends on two things - the type of gate it is, and the information fed to it.

There are three main types of gates, namely AND, OR & NOT. A refinement of the first two are the NOR OR, commonly called the NOR, and the NOT AND, commonly called the NAND.

As we said before, gates have two possible states. These states are known by many names, including on-off, true-false, and so on. The usual way of showing them is by the use of the figures 0 and 1. 0 corresponds to off or flase, and 1 corresponds to on or true.

An AND gate requires all inputs to be at 1 before the output will be at 1. Alternatively if any of the inputs to an OR gate are at 1 the output will be at 1.

We are not concerned with the other types of gates for this project. In passing, however, we might mention that a NOT gate merely complements an input. In other words, if the input to a NOT gate is 1, the output will be at 0, and vice versa. NAND and NOR gates are derived from the NOT gate.

If all inputs to a NAND gate are at 1, the output will be at 0, and vice versa. Similarly, for a NOR gate, if any input is at 1, the output will be at 0.

We will be using both the AND and OR gates in our alarm. The AND application will become obvious immediately. The OR application we will describe a little later.

The simplest concept of an AND gate lock is where a number of switches are connected in series and, by reason of this arrangement, in themselves constitute the AND gate. In place of the lamp we can use an electrically operated lock, a relay to control such a lock, or a power transistor to control a relay etc. Each arrangement would have its particular advantages.

If we use, say, four 12 position switches, with a random position selected on each switch, we have over 20,000 possible combinations, other than the right one.

In certain applications, such an arrangement can be very useful. For example, it can be used, not to operate a lock, but to disable an alarm system to permit legitimate entry. In such a situation, a would-be intruder has only one chance to get the combination right. If he fails the alarm will sound and he gets no second chance.

However, where such a circuit is used to operate a lock directly, the would-be intruder can have as many chances as he likes, the penalty for failure being nothing but wasted time. If he could work undisturbed, he would have a good chance of finding the combination by systematic trial and error.

The circuit which we have worked out below should be impossible to defeat. We say "should be" because nothing like this is absolutely 100% safe. However, the odds of someone cracking this combination are virtually infinite.

As well as a combination, the lock incorporates two other safety factors. If the wrong dial is moved first, an alarm will sound - so it is not enough to learn the combination. Secondly, even if the right dial is correctly set at the first attempt - which is very unlikely - the user is given only a short period in which to set all the other dials. If he takes longer than, say, 10 seconds, the unlocking mechanism is by-passed, and there is no way the lock can be made to open.

The period of setting time is best determined by the user. Anything from a few seconds to a few minutes is possible for this delay.

If the owner should accidentally take too long to complete the sequence he will, of course, be locked out also, but we have arranged matters so that the circuit will restore itself after a certain time, which can be selected by the user.

Now we can examine the circuit in more detail. The extreme left hand ("master") switch is wired differently from the other three. Once this switch is set the alarm is disabled, because the power for the alarm relay (RL2) and transistor (TR3) is supplied via the wiper and contacts of this switch. Once this is broken by setting the wiper to the correct position, the alarm is disabled.

Incidentally, while we have shown the master switch in a particular position, for reasons of circuit convenience, the actual position which it occupies relative to the other three, should be chosen by the constructor. Then he, and only he, will know which knob to operate first. Thus the position of this switch becomes a part of the total combination.

The other three switches are wired with their moving arms connected to an AND gate made up of three diodes, D1, D2 and D3. The combination number is connected to the 12V positive rail and the remaining contacts to the negative rail.

At this point readers may wonder why we have used diodes to provide the AND gate, rather than simply using the switches themselves. The answer is that while such an arrangement would be quite practical as far as this part of the circuit was concerned, it would prejudice the use of the additional protective circuitry we have already mentioned.

The operation of the diode AND gate is quite simple, and it will help the reader if he understands its operation. As shown in the circuit, it consists of three diodes with their anodes joined. This junction is the out-

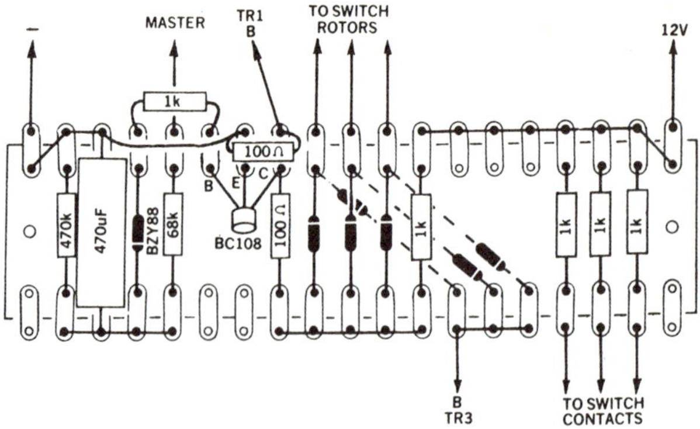

Component layout is not critical but that shown here is a logical one. The leads to relays, switches etc may be any convenient length.

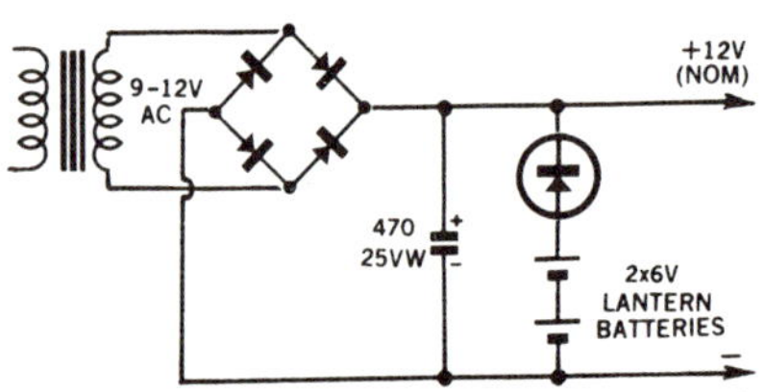

Basic circuit for a mains power supply with automatic battery standby.

put of the AND gate. The three cathodes are the three separate inputs to the gate. The output connects to the positive rail via a current limiting resistor, and also connects to the base of TR1.

In normal use (that is, with the switches set off their combination number) the cathodes of the three diodes are at negative rail potential. Therefore, they are forward biased and they conduct. Their anodes are only slightly positive with respect to the negative rail (the voltage across the diode), and the output is regarded as being in the 0 state.

As each switch is set to its combination number, the cathode of the associated diode is switched from the negative rail to the positive. Therefore, each diode ceases to be forward biased, and stops conducting. When the last switch is set, no diodes are conducting. The voltage at the output of the gate (anodes) rises substantially towards the positive rail voltage, and this shift constitutes the change from 0 to 1.

The increase in voltage is enough to turn on TR1, thus pulling in the relay. We used a 2N3055 power transistor in this project, rather than a BC108/relay combination which we have used in earlier projects. The main reason was that a number of readers have informed us that relays which pull in at low currents are relatively hard to come by. This is rectified by using a transistor which can handle more current.

The master switch circuitry has another function besides controlling the alarm. It also incorporates a time delay circuit, consisting of a resistor (R1), capacitor (C1) and a zener diode. It functions as follows: The resistor restricts the rate of current flow and the rate at which the capacitor charges. As the capacitor charges, the voltage across it rises until it reaches a point where the zener diode conducts.

This turns on TR2 (BC108), which "steals" the bias which would otherwise have allowed the power transistor to turn on and pull in the relay. Therefore, the lock cannot be opened after this time delay has occurred. If the legitimate user is caught by this trap, he has to wait until the capacitor discharges through its parallel resistor until he can try again.

The alarm circuit consists of TR3 (2N3055), relay RL2, three diodes D4, D5 and D6, plus a few minor components. TR3 is connected between the positive rail, via the master switch, and the negative rail, with RL2 in the emitter circuit. The base of TR3 connects via a suitable limiting resistor to the output of an OR gate made up from the three diodes.

These three diodes are connected to the three combination switches in such a manner that, if the moving arm of any switch is set on, or passes over, its combination number, the positive rail voltage is applied to the diode, which becomes forward biased, and so to the base of TR3. This turns TR3 on, the relay closes, and the appropriate set of contacts operates the alarm. A second set of contacts between the positive rail and the base of TR3 by-passes both the switch and TR3, thus keeping the bell ringing despite any manipulation of the switches.

Note that only one diode needs to be forward biased in order to turn on TR3. This can be either D4 or D5 or D6; hence the term OR gate.

Now that we have explained the electronics of the combination lock, we can have a closer look at the mechanical aspects. Naturally enough, the switch mechanism represents the most important part of the device. We understand that a number of manufacturers market switches specifically intended for purposes such as ours. These are known as "thumbwheel" switches, and are operated by turning a numbered wheel which is mounted edge-on through a facade. It is a simple matter to operate these, and select the correct number very quickly.

For those who are more interested in building the combination lock as a novelty, we recommend conventional rotary switches, single pole - 12 position. These can be a miniature type.

Ideally, these switches should have non-shorting contacts (i. e. break-before-make) to avoid complications when switching between contacts at different voltage levels.

Electrically operated locks may be new to most readers, but they are readily available. The well known Yale company market several models.

The bolt is normally spring loaded in the locked position, and is withdrawn when power is applied, and only for as long as it is applied. They are available for 12V operation and require a current of between 800mA and 1A.

The power supply required for the whole system depends on a number of factors. The basic circuit, as shown, can be operated from a relatively simple supply such as a pair of lantern batteries. In the locked position there is a steady drain of about 12mA and in either the unlock or alarm position the drain will depend mainly on the resistance of the relays used in these circuits. Assuming typical relays of about 500 ohms the operate drain for either combination would be about 25mA.

Power for the associated devices, such as the alarm or power operated lock, presents more of a problem. A heavy duty alarm may require 1A or more and need to operate for a reasonable period if it is to be effective. As already stated, the lock will require a similar current, but with a shorter duty cycle.

This kind of power can be supplied by dry batteries, and the two lantern batteries already mentioned may be adequate in some cases. However, these would be right on the lower limit.

Mains power is an alternative, but has some objections. One is that an unauthorised person may attempt to beat the system by cutting off the power. This will, in fact prevent the lock from operating, but it will also disable the alarm, which may be undesirable in some circumstances. Another objection is that a power failure may prevent legitimate entry, although some locks have a mechanical override using a conventional key.

A good compromise appears to be mains operation with battery standby. This is quite easy to arrange and has a high order of reliability. As shown in the circuit, the mains supply consists of a transformer and rectifier delivering a little more than 12V. The standby battery is permanently connected, but isolated from the load by means of a diode, which is reversed biased so long as the mains supply voltage is higher than the battery voltage. If the mains supply fails the diode is forward biased and the battery takes over without any break.

Adjustment of the time delay circuitry is simple. The 68K series resistor gives a time delay of around thirty seconds. To shorten the delay, one of two changes can be made. The 68K resistor can be replaced with a lower value - a 22K should give around ten seconds delay. Alternatively, the zener diode could be replaced with one of lower voltage. A lower voltage zener does not require the capacitor to charge as long to reach the zener point.

Using the lock is very simple. However, always make sure the master is the last switch un-set, otherwise the alarm will sound. Also, it must always be the first switch set, for the same reason. Remember the time delay, too. If the correct combination is set, and the lock does not operate, un-set all switches (the master last) and wait for about thirty seconds. This gives the capacitor time to discharge. Then, the setting can be commenced again.

PARTS LIST

TRANSISTORS etc.

2 2N3055 or similar
1 BC108
8 Diodes BA100
1 BZY88/C7V5 zener diode

RESISTORS ($\frac{1}{2}$ W)
1 470k
1 68k
5 1k
2 100 ohms

CAPACITORS

1 470uF 16VW electrolytic

MISCELLANEOUS

1 Relay, 100 to 500 ohm coil
1 Set normally open contacts
1 Relay, 100 to 500 ohm coil
2 Sets normally open contacts
4 Rotary or thumbwheel switches, see text.
Suitable alarm (if required)
Suitable lock (if required)
Tagstrip, hookup wire, solder etc

ADDITIONAL PARTS FOR POWER SUPPLY (Optional)

1 power transformer, 240V to 12V 1A
5 BA100 diodes
1 470uF 25VW electrolytic
2 6V Lantern batteries

THE VARI LIGHT

HAS YOUR WIFE EVER SAID TO YOU, 'I WOULD LIKE TO DIM THE ROOM LIGHTS SOMETIMES". IF SO, THEN HERE IS THE VARI-LIGHT, AN UNOBTRUSIVE WALL MOUNTING CONTROL TO MAKE A ROOM BRIGHT OR DIM - OR ANY LEVEL IN BETWEEN.

There are many occasions in the home when the full brilliance of the lights is not required; when watching television, listening to music or just plain relaxing and, of course, at parties when the lights can be dimmed "way down low". A light dimmer such as the Vari-light can be used to advantage in bedrooms so that lights may be left on at low level; a particularly useful facility when children are sick or for those who are nervous in a completely darkened room. If for these, or any other reasons, you wish to control domestic light levels, the Vari-light is the answer.

Externally, the Vari-light appears as a mains power point surround with a blank face and a knob in the centre. Inside, it contains a standard Triac light dimmer circuit which will control incandescent lamp loads up to 300 watts.

For readers who may be unfamiliar with the operating principles of light dimmers using Triacs, the following explanation should be helpful. The heart of the Vari-light is a Triac, a silicon AC control device, originally developed by General Electric of U.S.A. It is closely related to the thyristor and is triggered into conduction by a gate signal in the same way as the thyristor. For the purpose of this explanation, a Triac can be considered as a pair of thyristors connected in inverse parallel with a common case and common gate electrode.

In essence, the Triac is a bidirectional switch which after being triggered into conduction, stays "on" until the supply voltage decreases to zero or reverses in polarity, when it turns off and can be switched on again. Used with AC, a Triac can be triggered into conduction at any point on either half cycle by a low voltage signal of either polarity applied between the gate electrode and terminal 1 (anode 1). Note that since the Triac is a bidirectional device, it has no anode or cathode as such. The two end terminals are normally referred to as "anode 1" and "anode 2" or "terminal 1" and "terminal 2".

As the Triac is a switching device which is either fully conducting or "off", the only means by which it can be used to obtain variable control of power is to use it as a very rapid switch which closes for variable periods of time during each half-cycle of the AC voltage waveform - by adjusting the instant during the half-cycle when it triggers into conduction.

While there are many methods of varying the triggering point of a Triac, the most satisfactory one is known as "phase control". This involves applying to the gate electrode a sharp pulse of current whose phase, relative to the AC waveform, can be varied. This is done by means of a capacitor which is charged while the Triac is in the non-conducting state. The time the capacitor takes to charge will depend on its size, the resistance in series with it and the voltage supplied to it.

Figure 1 shows a basic Triac light control circuit employing phase control. In this circuit the capacitor and the charging voltage (240VAC) are constant and the triggering point of the Triac is varied by adjusting the resistance of the rheostat. The capacitor's charge is delivered to the gate of the Triac via a voltage sensitive device which conducts only when the voltage across it reaches a certain value. The voltage sensitive device used here is called a Diac, a three-layer symmetrical breakdown diode which is an open-circuit until the applied voltage rises to breakdown rating, whereupon it breaks down to a negative resistance. The breakdown voltage for Diacs is generally of the order of 30V in either direction, i.e. they are bi-directional devices.

Figure 2 shows the waveform that is applied to the lamp at different power levels. The dotted sine wave represents the mains voltage. Figure 2(a) shows the Triac firming late in each half-cycle which corresponds to a low level of power. Similarly, figure 2(b) shows the waveform at half-power operation and figure 2(c) shows the waveform at a high power level where the Triac is firing early in each half-cycle.

Although a simple light dimmer can be constructed using the components shown in figure 1, additional components are added to this to reduce hysteresis effects, extend the effective control range of the rheostat, and suppress radio-frequency interference.

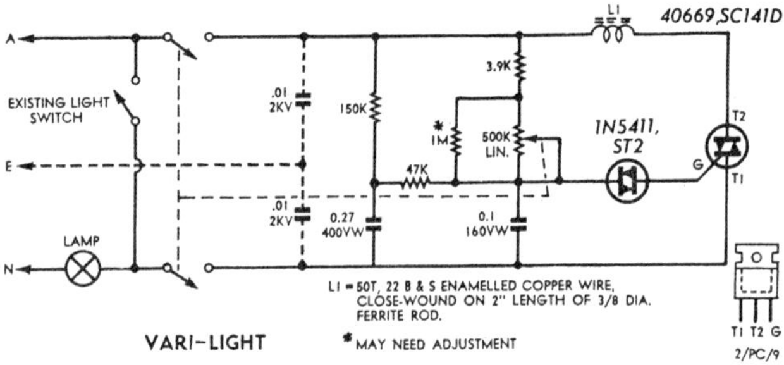

The main circuit, showing its connection to the existing lighting circuit. The 1M resistor should be adjusted to give the greatest range of control. Additional interference suppression wiring and components are shown dotted.

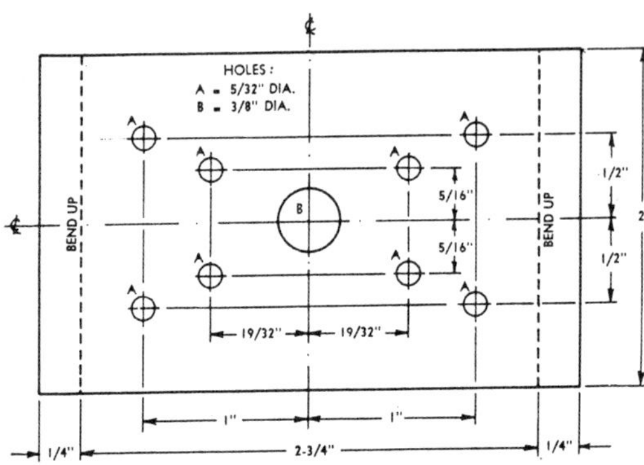

Dimensions and hole positions for the metal plate on which the components are assembled. Aluminium in 22g or 24g would be a suitable material. It is mounted on the blank face-plate.

When applied to light dimmers, the term hystersis refers to a difference in the rheostat setting at which the light initially turns on and the setting at which it is extinguished. With the simple circuit in figure 1 the rheostat may have to be turned through 30 to 40 per cent of its rotation before the lamp begins to glow and at this initial setting the lamp will be quite bright. The rheostat has to be turned back to a much lower setting to extinguish the lamp.

Besides giving a poor control action, hysteresis is undesirable because, at low levels of illumination, the light may be extinguished by a momentary drop in the mains voltage; for example, when a radiator, washing machine or other high-power domestic appliance is switched on.

Hysteresis is caused by the sudden drop in capacitor voltage when triggering first occurs. Figure 3 shows the charging cycle of the capacitor. The large sine wave shows the mains voltage (240V AC) while the discontinuous waveform shows the voltage across the capacitor. The two waveforms are not drawn to the same scale. The initial triggering of the Diac occurs late in the first half-cycle of the mains voltage, causing an abrupt drop in the capacitor voltage just after the capacitor voltage rises to Vbo, the Diac breakdown voltage. As a result, the capacitor recharges in the opposite direction to the Diac breakdown voltage sooner than if the abrupt voltage drop had no occurred. This means that the Diac fires earlier in the following half-cycles and so the Triac delivers more power to the lamps than is desired for an initial setting.

The hysteresis effect can be reduced if the decrease in voltage across the capacitor, caused by the Diac firing, can be minimised. This can be done by feeding the capacitor from another RC network, forming a "double-time constant" network. The added capacitor reduces hysteresis by charging to a higher voltage and maintaining voltage on the capacitor which is being discharged by the Diac.

The ideal situation, giving maximum range of light control, is obtained when the lamp begins to light as soon as the control is turned slightly away from the zero setting. The addition of the "double-time constant" network to the circuit improves the situation but initial triggering may still occur at about 40 per cent of the control range, meaning that the actual brightness control is only effective for 60 per cent of the control rotation. Performance can be improved still further by the addition of a resistor in parallel with the rheostat, the value being selected experimentally to give the maximum control range. A 1M resistor was needed in the prototype, but may vary with individual units.

With a "double time-constant" circuit and a resistor across the rheostat to adjust for maximum control range, the result is a dimmer with very smooth control over just about the whole range of the rheostat. The purpose of the 3.9K resistor in series with the rheostat is to protect the resistance element by limiting the capacitor charging current when the rheostat is at low resistance settings.

In operation, the Triac switches on in about 1 or 2 microseconds, the current rising to whatever the load permits within this period. The rapid rise in current produces radio interference extending up into the region of several megahertz. This will affect the broadcast band and the lower short-wave bands. The interference will be evident as a buzzing noise.

RF interference caused by the Vari-light is reduced by the inductor, L1, which is wound on a ferrite rod. In most light dimmer circuits, several .01uF suppression capacitors are connected between active points in the circuit and earth. However, since the Vari-light will be used in the lighting circuit of homes no earth wire will be available for these suppression capacitors. Because of the 300 watt limit we have placed on the dimmer the current level is not great and consequently the interference is only moderate - less than that radiated by most commutator motors in domestic appliances such as food mixers.

If radio reception in your area is weak the RF interference generated by the Vari-light may be unsatisfactory. The RF suppression which is satisfactory in strong signal areas will have to be improved upon. One method of improving RF suppression is to connect 0.01uF/2KV ceramic disc capacitors from both sides of the Triac to an earth wire, as shown by the dotted portion of the circuit. The whole of the dimmer circuitry can be installed in an earthed metal box to shield the inductor. Finally, all the lighting cables could be run in earthed metal conduit as the ultimate solution.

The Triac used is the economy, plastic-encapsulated unit, 40669, made by RCA. It has ratings such that it can control a load of up to 300 watts without the need for any external heatsink. The Diac used is also made by RCA, type 1N5411.

The Vari-light is encased in a standard mains power point surround with a blank face-plate. These are readily available from electrical supply and hardware stores. These components themselves are mounted on a metal plate which, in turn, is secured to the face-plate by means of four $\frac{1}{4}$ in long, 1/8 in Whitworth screws in ready-tapped holes.

No electrical connections are made to the metal plate, for safety reasons. All terminations are made to two tagstrips one at each end of the plate and the 500K potentiometer which is wired to function as a rheostat.

The metal plate is 2in wide with an overall length of 2-7/8in, including a 2in. flange at each end. The accompanying diagram shows the dimensions and positions of the holes which should be checked against the tapped holes in faceplate. The thickness of metal is non-critical - 22 or 24 gauge would be a good choice.

The interference suppression inductor L1 is not available commercially but is quite easily made. Start by winding a layer or two of thin insulation tape on a 2in length of 3/8in diameter ferrite rod. If a full length

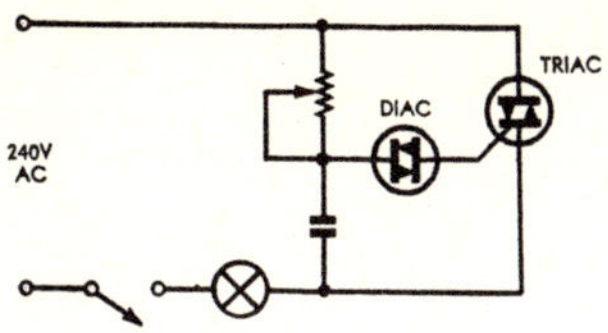

Figure 1

A basic Triac control circuit which serves to illustrate the principles involved.

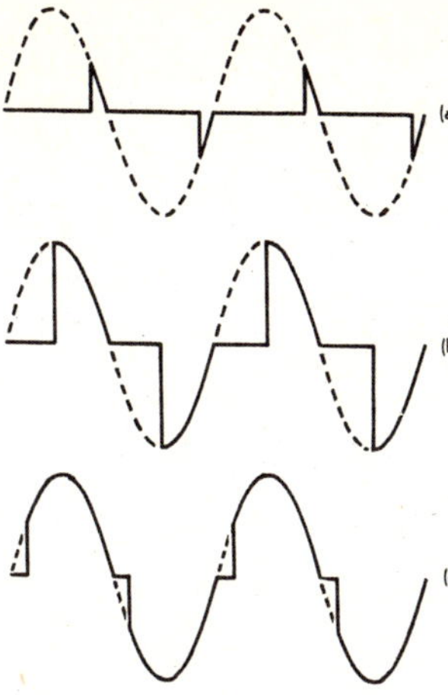

Figure 2

Waveforms applied to the lamp at various power levels. The dotted line shows the mains voltage, the solid line that applied to the load.

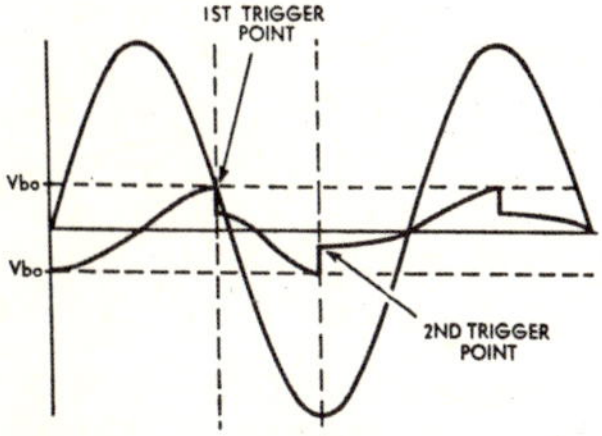

Figure 3

Waveform showing the cause of hysteresis: the drop in voltage across the capacitor when the Triac fires.

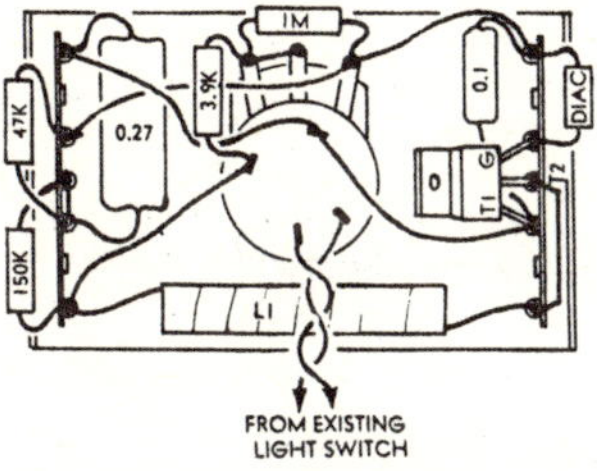

Wiring diagram showing the approximate disposition of the components.

ferrite rod has been purchased it can be cut by filing a nick right around the circumference of the rod and then snapping it as if it were of glass - do not try to saw the rod. Close wind 50 turns of 22 gauge enamelled wire over the insulation tape. Then wind insulation tape tightly over the rod in a couple of layers. This last step is important - if it is not wound tightly the inductor will make a buzzing noise due to currents being switched by the Triac.

The capacitors used in the prototype are dipped polyester types. The recommended voltage ratings must be adhered to. If they are low the capacitors will break down; if they are high they will be too large for the available space.

The order of assembly should be as follows: first, cut the potentiometer shaft to a length of ½in. Attach the potentiometer and two tagstrips to the metal plate. Drill a 3/8in hole in the centre of the faceplate. Care is needed here, otherwise the plate will fracture. Use three drills to bring the hole to size, e.g. 1/8in, ¼in and 3/8in. Then attach the metal plate to the faceplate by means of ¼in long, 1/8in Whitworth roundhead screws.

The components can now be soldered into place. The two capacitors and Triac should be installed first. The Triac leads should be trimmed to ¼in and bent slightly apart. Take care not to fracture the leads as they are bent. After soldering the capacitors and Triac into place, the Diac and resistors can be installed. The 3.9K resistor should have a spaghetti sleeve fitted to the lead which is soldered to the potentiometer switch, to avoid shorting the switch to the potentiometer case. Take care to ensure that no connections are made to the mount-feet of the tagstrips.

Lastly, the Triac should be wrapped with one or two turns of insulation tape to ensure that the flag does not short to the potentiometer case. Ensure that all the components are within the boundary of the metal plate so that they do not foul the power point surround when the faceplate is being installed.

A plastic knob should be used on the Vari-light so that the metal plate and potentiometer case are completely isolated. The knob should be installed so that its skirt is almost flush with the surface of the faceplate.

Having checked the wiring against the wiring and circuit diagrams, final testing and adjustment may be carried out. This is best done by wiring the Vari-light to a three pin-mains plug and connecting a light socket in series with the active lead, as shown in the circuit diagram. The test lamp may be the same as that to be used, although this is not critical. The brightness should increase smoothly with rotation of the knob. Adjust the 1M resistor for the desired initial level of brightness.

Since this is a mains operated device it may be wise to seek the advice of an electrician before attempting to install the unit, or even have him make the actual connections. In any case, the appropriate light fuse must be removed and/or the main switch turned off before any work is done on the lighting circuit. As a further precaution, make sure that the light switch is left in the "on" position and that the lamp is functioning, when the power is turned off at the switchboard. This will provide an indication that the correct circuit has been opened and also discharge any capacitors, such as fluorescent power factor correction capacitors across the line. Such capacitors can deliver a frightening, though relatively harmless, shock in some circumstances.

There are several ways in which the unit may be installed. Perhaps the most satisfactory arrangement is to connect it in parallel with the existing light switch - this means that the light switch will override the Vari-light so that a person entering a darkened room can light his way without fumbling with knobs. In this case, installation would be a matter of securing the power point surround to the wall with suitable screws, taking two wires from the potentiometer switch and soldering them to the existing light switch.

The Vari-light should be fitted with a pair of flying leads of suitable length to run behind the architrave, through the holes from which the existing switch wires emerge, and into the switch terminals, where they will be screwed down alongside the regular wires. To facilitate feeding the flying leads through the holes already occupied by the switch wires it may be necessary to enlarge these holes and feed "fish wires" through them from the front. These are then joined to the flying leads and used to pull the latter through the holes from behind.

In this way it should be possible to complete the operation without the need to use a soldering iron while the power is turned off. Alternatively, it may be possible to use the type of insulated screw connectors which electricians use in junction boxes, again avoiding the need for a soldering iron. If soldering cannot be avoided it is usually possible to leave the power circuit energised while the light circuit is switched off, since these are separate in most houses. However, be very careful that these two circuits are not confused.

Remember that the maximum load to be controlled by the Vari-light must not exceed 300 watts - at this power level the Triac will be dissipating approximately 1.25 watts. With more elaborate heatsink arrangements the same circuit is suitable for loads up to 1,000 watts. Fluorescent lamps cannot be controlled with the Vari-light.

PARTS LIST

1 Mains power point surround and blank faceplate
1 Metal plate to suit faceplate.
1 Ferrite rod inductor (see text)

1 500K (lin) potentiometer with rotary DPST switch
1 Plastic knob
1 Triac, 40669
1 Diac, 1N5411
2 7-lug tagstrips (with two mounting feet)

RESISTORS ($\frac{1}{2}$ watt rating)

1 x 1M (see text), 1 x 150K, 1 x 47K, 1 x 3.9K

CAPACITORS

1 x 0.27uF /400VW polyester
1 x 0.1uF /160VW polyester
2 x 0.01uF /2KV ceramic disc

NOUGHTS AND CROSSES

AN ATTEMPT TO DESIGN AN UNBEATABLE NOUGHTS AND CROSSES MACHINES

The author claims that the machine cannot be beaten, but will always win or draw against logical play. The chief disadvantage is that the machine must always go first, taking the bottom left-hand square. The game is played on a piece of paper with the pattern drawn in the usual way and with an X (for the machine) in the bottom left-hand corner. After each move by the opposing player, the machine indicates the move to be recorded on its behalf by a lamp lighting in the appropriate square.

With the machine already occupying square 7 in the bottom left corner the player makes his first move by writing a nought (0) in the square he selects. He then sets switches S2 to S5 according to the combination adjacent to the lamp on the panel in the chosen square. For example, if the player wanted to place his 0 in the top right-hand corner (square 3), he would set S2 to 1, S3 to 1, S4 to 2 and S5 to 0, although the position of S5 is of no account in this example. The player then rotates switch S1 one position, the lamp in square 7 (indicating the machine's initial move) goes out, and another lamp will light to indicate the machine's next move. This is written (X) on the written diagram on the paper.

The player makes his next and succeeding moves writing a nought in the appropriate square and operating the eqivalent toggle on the panel of toggle switches. Switches are fitted in positions 4, 5, 6 and 8 with dummy switches in the other positions. After making his second move by depressing the toggle switch corresponding to the selected square, the player rotates switch S1 another position to discover the machine's next move. This procedure is repeated using the toggle switches on this panel until the game is completed. While some of the toggles are dummies, the guest player need not know this.

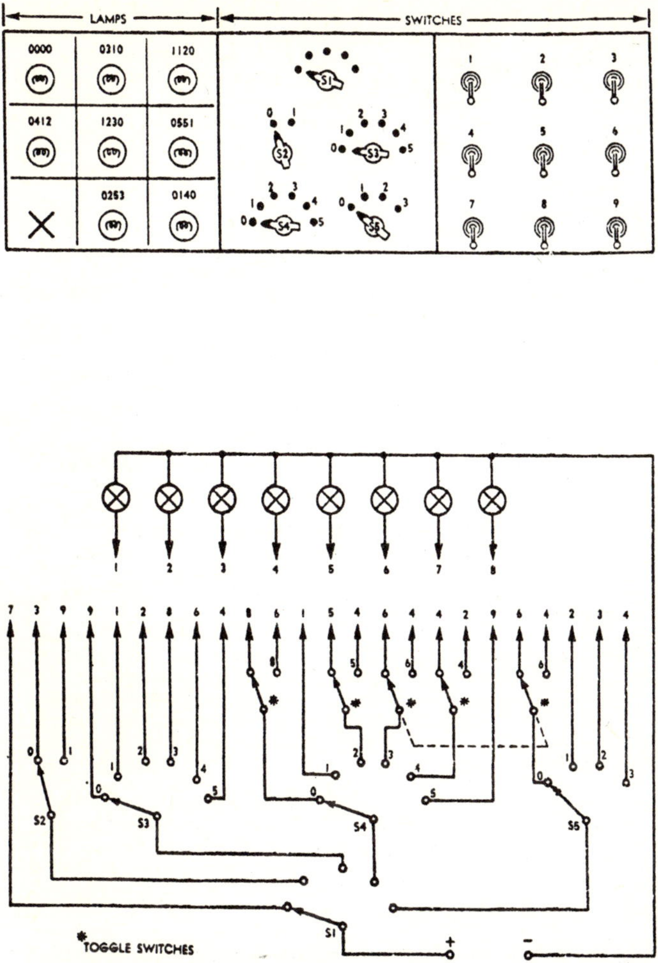

TOP: A possible panel arrangement for the noughts and crosses machine. BOTTOM: The circuit of this simple device.

Although the machine always wins or draws against logical play, it is unable to copy with illogical moves by the player. The "program" assumes that in every case the player will make the obvious move to prevent a row of Xs. If the player does not make this move by oversight or by deliberate tactics, the machine is not capable of sensing the possible alternative "win" pattern. Following the choice of certain initial blocking moves by the player, the machine will always play for a draw oblivious of any winning opportunities which may occur. If a player makes one specific initial move followed by one of two illogical moves ignoring a threat of a row of Xs, the machine will allow the player to complete a row of 0s because it cannot foresee such an action by the player! As soon as the way to beat the machine is discovered, its amusement value will decrease rapidly. More sophisticated designs which really are infallible have been developed and may typically be seen at fetes etc.

PARLOUR GAME

DESCRIBES AN ELECTRONIC VERSION OF A POPULAR PARLOUR GAME.

This circuit is of a very simple "computer" suitable for construction by younger readers. It will play the following game: Twenty draughts (or other objects) are laid out in a row and each player takes turns to remove one, two or three draughts. The object of the game is to force one's opponent to take the last draught.

It can be constructed in a small box with nails projecting from the top or front and wired together underneath as shown in the diagram. The crocodile clip is moved by the player, according to his own inclination when it is his turn, or according to the instruction issued by the computer when the "computer's turn" button is pressed. This is a simple loaded press button.

To start the game, the clip is attached to pin 20, indicating that there are 20 draughts still to be removed. A decision is then made - perhaps by tossing a coin - as to whether the player or the machine has first move. If the decision favours the player, he moves according to his inclination. If it favours the machine, the "computer's turn" button is pressed and the instruction acted upon.

In fact, the machine is unbeatable if it has first turn. It will usually win if the player has first turn, but it can be beaten.

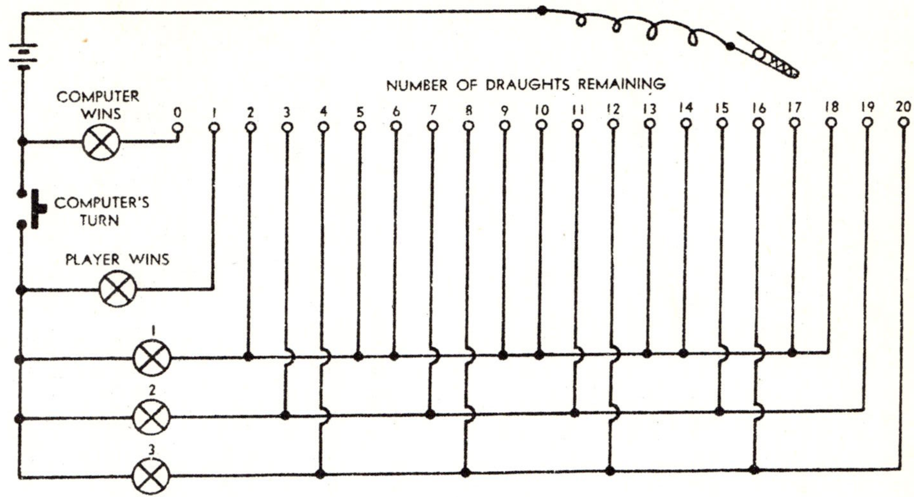

Circuit of Popular Parlour Game

DECISION MAKER

Operation of this circuit may be considered as analogous to the tossing of a coin. Most of the circuit components used were from computer boards, although other readily available components could also be used. The circuit is relatively simple.

The monostable consisting of TR1 and TR2 is triggered by S1 and causes TR3 to conduct for the period of the monostable, the times being approximately 1 second. This in turn switches on the astable multivibrator consisting of TR4 and TR5 and during this period, the number of output pulses varies somewhat according to transistor cutoffs etc. The output pulses are fed to a flip-flop which performs a divide-by-two counting function with a light emitting diode in the collector of one of the transistors. This gives an indication of the final state of the flip-flop. If an even number of pulses are fed into the flip-flop, then the output will revert to its original state. If an odd number of pulses are fed in, then the opposite will be true and hence the equivalent to a heads-tails function.

During activation, the LED will appear partially illuminated but will actually be switching on and off rapidly. If the light output is unsatisfactory, the collector of the flip-flop may be reduced equally. If the read out function is not random enough, the speed of the astable multivibrator may be increased by reducing the value of the two 62K resistors.

While we are aware that there are simpler ways of doing the above, the exercise involves three different types of multivibrator and thus could provide valuable experience for students and experimenters.

NEW PARLOUR GAME

ALTHOUGH THIS IS YET ANOTHER VERSION OF THE ELECTRONIC GAME DESCRIBED PREVIOUSLY, IT OFFERS SOME NOVEL FEATURES. ONE IS THE METHOD OF DISPLAYING THE "MACHINE'S MOVE" INSTRUCTIONS AND ANOTHER A MEANS TO VARY THE MACHINE'S "PROGRAM" WHEN THE PLAYERS LEARN HOW TO BEAT IT.

The idea is not new but readers may welcome a switching circuit which can be used to entertain one's friends.

To play the game 14 switches are turned down in sequence from left to right. The player and the box take it in turn to move, turning down one, two, or three switches at each move, as they please. Each tries to leave the last move to the other. When it is the box's turn to move the player presses the switch S18 and then, on the box's behalf, turns down the switches over which a light appears. It is then the player's turn again.

Decision Maker

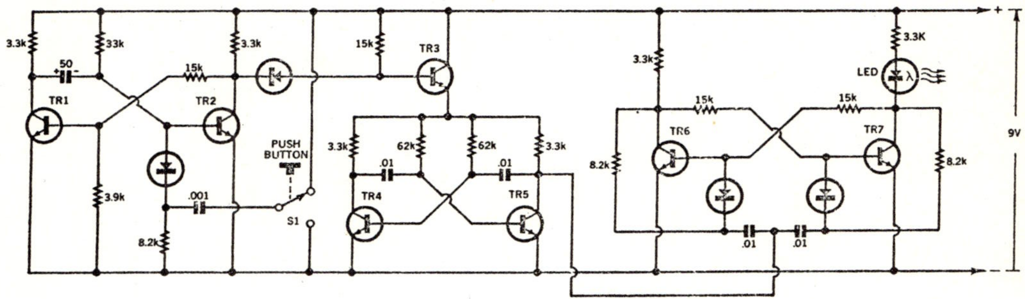

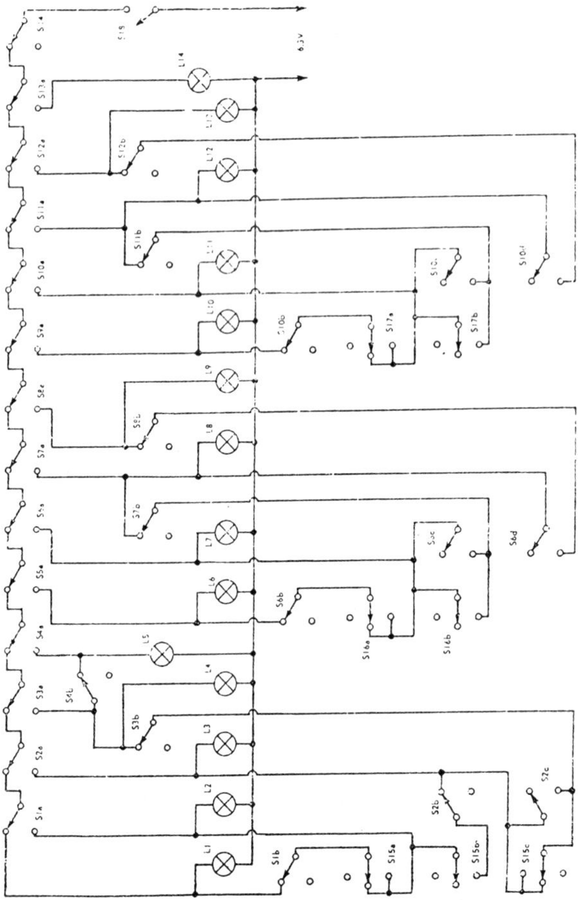

Circuit of New Parlour Game

The switches S15, 16, and 17 are used to vary the circuit when someone beats the box, so that he cannot win again by making the same moves.

I purchased 18 obsolete Post Office lever keys. Switches S1 to S14 are two-position types. Most of them need to provide three change over functions (the "a" "b", and "c" functions shown on the circuit) and two, S6 and S10, need four such functions ("d"). These switches are normally set in the horizontal position (the position represented on the circuit) and pushed down to represent the various "moves". Switches S15 to S17 are three-position types, and S18 is a spring return type.

I mounted the switches in a wooden frame, arranged in two rows. From S1 to S14 makes up the top row and S15 to S18 are spaced out to make the bottom row. There is a 7/16in hole above each switch S1 to S14 to accommodate a 6.3V indicator bulb with glass cover.

IMPROVED VERSION OF PARLOUR GAME

WORKING OUT ELECTRONIC PARLOUR GAMES APPEARS TO BE A POPULAR PASTIME AT THE MOMENT.

This circuit is mainly a modification of the electronic parlour game.

As stated in that article, the game is played with 20 draughts or other suitable objects. Each player takes turn at removing either 1, 2 or 3 draughts, the object being to force one's opponent to take the last draught.

The original game was constructed in a wooden box with a row of nails protruding from the lid, numbered 0 to 20. The nails were part of the circuit and a crocodile clip on a wander lead was placed on the numbered nail corresponding to the number of draughts remaining, after the player or machine made their alternate moves.

I constructed this game and found it quite satisfactory except that confusion sometimes arose when a player forgot to advance the crocodile clip - mainly on the machine's behalf. Recently, I decided to build a version which would be free from this problem.

The basis of my design is a uniselector: a type of rotary switch with four or more banks and 25 positions on each bank. The rotor arm is advanced one position every time a pulse is fed to an electro-magnet in the device. The other components are a dial mechanism, a key switch, and a number of lamps and sockets.

A player makes his move by dialling either 1, 2 or 3 on the dial, which advances the uniselector the corresponding number of positions. The machine's decision and play is initiated by operating the key switch,

Improved Version of Parlour Game

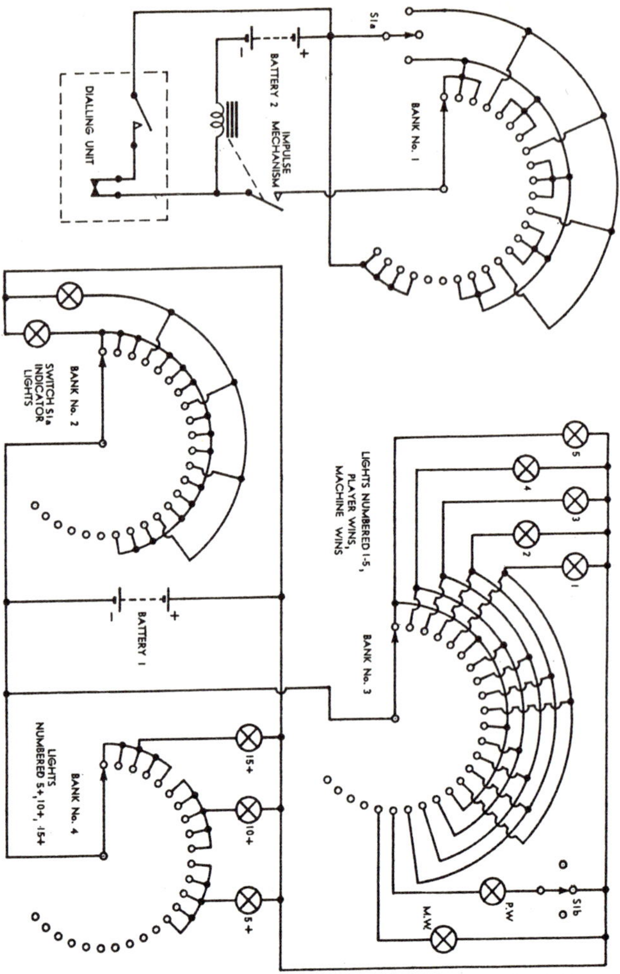

A novel feature of this circuit is the method by which only eight lamps are used to count up to 20. Note that a power supply can be substituted for battery 2.

S1a. This is normally in the mid position but is moved to either the right or left-hand position when it is the machine's turn to play. The choice of right or left-hand position is indicated by one of two lights (connected to bank No. 2) which are mounted on each side of the switch.

An indicator board is provided which reads off the number of draughts remaining. Instead of 20 lamps numbered 1 to 20, I have reduced the number to eight, labelled as shown in the diagram. For example, 17 would be represented by lamps number 15 and 2 both being illuminated at the same time.

Two more lamps are provided (connected to bank No. 3) one to signify that the player has won and the other that the machine has won.

The battery connected to banks 2, 3 and 4 has only to light the various lamps and so its voltage will be chosen accordingly. The battery on bank No. 1 is required to operate the uniselector mechanism and should be between 24 and 50 volts. I used an old 45-volt "B" battery, but plan to replace this with a power supply.

As with the original version of the game, the choice of first turn can be determined by the toss of a coin. If the machine has first turn it cannot be beaten, but it can be beaten if the player has first turn, although this is not easy.

The key switch S1 has two function. S1a initiates the machine's turn, as already explained. S1b is associated with bank No. 3 and operates the "Player Wins" light.

I constructed my game in a case with a sloping front. It is finished with contact covering and with the addition of coloured bulbs, the whole thing looks very impressive.

SIMPLE SIGNALLING DEVICE

This circuit was set up between workshop and kitchen so that a signal could be given to indicate that tea was ready, etc. Since the signal could sometimes be drowned out by the noise of machine tools, the two-way facility was needed to allow the person in the workshop to indicate that he had heard the summons. Advantages of the circuit are that it requires only two conductors between stations, needs a power supply at one station only and is very inexpensive.

The diodes are arranged so that normally no current flows. When push-button "A" is closed, the positive half-cycles from the transformer can pass to bell "B" and cause it to ring. When push-button "B" is closed, the negative half-cycle caused bell "A" to ring.

Power was taken from a low voltage bell transformer already in use for the front door bell. Figure of eight flex was used to link the two stations. The diodes used were 1A diodes. With the assistance of parts from the junk box, the final cost was little more than that of the four diodes.

Simple Signalling Device

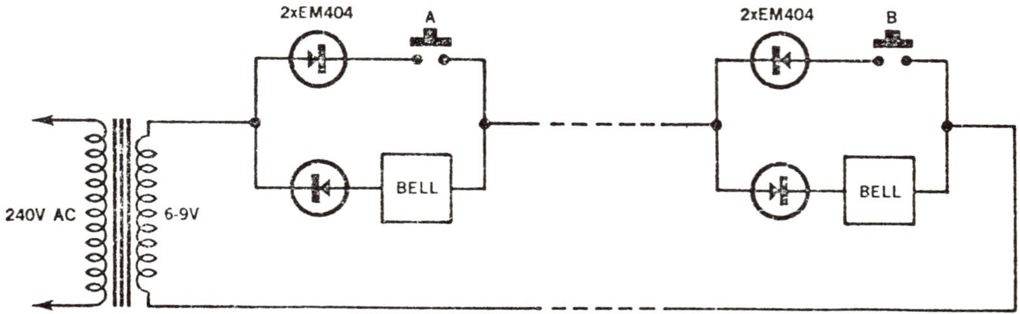

shunted by this low reactance (now essentially the 50uF capacitor C1) that oscillation can no longer be sustained.

The AC128 is now turned on so hard that a change-over pulse through the .0033uF collector coupling capacitor to the base of the AC125 (or AC126) is generated. This turns this flip-flop over, completely cutting off the AC128 via the 50uF coupling capacitor, C2.

The time constant of the 22K/2uF series RC circuit now takes over, and holds the flip-flop over for the period between "chirps". This then allows the AC128 to switch on again to re-establish oscillation. During the cutoff period of the AC128, the 50uF capacitor C1, loses some of its initial charge, but regains it during the next "on" period. The result is that the charging cycle of this capacitor is repeated for a short time, again passing current through the inductor to produce the desired effect of a fast, changing "chirp" frequency.

When the long period flip-flop changes over, the AC128 is cut off completely, preventing any of the foregoing actions from taking place. The 50uF capacitor C1, then loses almost all its charge, with the entire process just discussed repeating itself after the long period flip-flop changes back.

Having analysed the behaviour of the circuit we then considered how practical it would be to duplicate it with components readily available on the local market. The biggest problem was the two miniature transformers involved. We had to consider whether we could find a local product with the same characteristics, or evolve a circuit which would achieve the same end result without recourse to transformers at all.

While awaiting delivery of sample transformers which we felt might be suitable, we investigated this other possibility, using flip-flops and sawtooth generators. After a lot of juggling with time constants and mark space ratios, we achieved a result which was at least promising, even though it was a long way from what was required. However, by this time the flip-flops controlling the "chirp" and "on" period of the sawtooth oscillator had grown in complexity and number of components to an uneconomical degree. Our bird appeared to be taking on the size of a vulture rather than a canary!

Fortunately, subsequent experiments with locally available transformers indicated that they could be substituted in the original circuit. The local units are physically much larger than the original ones, but are otherwise very suitable requiring only a few changes of resistance and capacitance around the circuit to produce a result very similar to that provided by the original.

There is plenty of room for experiment here. Once the basic functions have been achieved one can fiddle with circuit values to one's heart's content in pursuit of the elusive "perfection". In fact, we present the idea as an electronic exercise and basis for experiments, rather than a serous project in its own right. Considering the relatively small cost of the components involved, we feel that this approach is justified.

We deliberately chose the cheaper type germanium transistors for this project in order to keep cost to a minium. They are quite adequate for the job and temperature changes are not a serious problem. A further saving was made by assembling the whole unit on a small piece of Veroboard, rather than involve the builder in the unnecessary expense of a printed wiring board for such a simple configuration. A drawing of the layout of this board is shown and is based on 0.15in between holes.

The speaker used is a small 8 ohm unit, and it can be either a 2" or 3" one. The circuit board and speaker can be mounted in an imitation birdcage or a case of strictly utilitarian appearance. This is, however, left to the fancy of the individual reader.

A suitable housing for the "Hot Canary" electronics can be made from virtually any decorative plastic box of sufficient size to take all components. A visit to the local toy shop or department store should provide plenty of inspiration, limited only by one's imagination and how far the finished product needs to be "prettied up". We made our bird's environment from a small plastic electrical junction box.

Two countersunk screws passing through the bottom of the box are used to mount the Veroboard assembly, spacing this off the box by one nut thickness, and clamping with a further nut on the topside of the board at each screw.

The power supply is a small nine volt radio battery such as the type 216. The battery position is determined by the layout of other components within the box. It can be held in place by a simple bracket fashioned from an aluminium strip fastened with a short screw and nut through the bottom of the box. The ON/OFF switch can be any suitable unit such as a slider or push-push type. The layout inside the box is not critical.

In gluing the cloth, the panel was shaped so that it fitted the inside rim of the box, and the speaker hole cut in a position determined by internal layout of the battery and Veroboard circuit board.

Miniature speakers of the size used in this project are not usually provided with mounting holes in their rims. As installed in portable radios, they are usually located in a part of the injection moulded cabinet by two or more clamping lugs.

We overcame this mounting problem very easily by first placing the speaker in the desired position and marking its outline on the panel with a pencil. From each side and the end nearest the speaker, three lugs were formed by cutting with a pair of snips from the outside edges towards and up to this pencil line. The lugs were then bent up at the pencil line and trimmed to sufficient length to enable them to be bent back over the rim of the speaker to secure it reliably. The method used can be clearly seen in the photograph of the speaker on its panel.

If the system exhibits a noticeable "plop" at the beginning and end of each warble this may be minimised by placing adhesive tape over the openings at the rear of the loudspeaker. This will increase the back loading of the cone, thus raising the resonance.

A simple carrying handle made from two hoops of plated steel wire form a simple "cage" for the canary and a convenient mounting place for his perch. Sprigs of imitation leaves (and flowers if you like) take away the bareness of the bird's environment.

Possible sources of birds and foliage are those companies who specialise in display materials. These are usually listed in the Classified telephone directory under "Display and Exhibition Equipment and Supplies". A local department store or toy shop may also have suitable items. The bird is made of papier mache with real tail feathers. Its legs and claws are plastic which are extended into thin steel wire for the purpose of mounting. The foliage is made of plastic.

Fairly obviously, there is almost no limit to the degree of realism which one may strive for in such a display. Real cages, more life like birds - even real ones (stuffed) if you can afford it - and with as many trimmings as you like are all possible if you feel that the realism of the song justifies it.

In any case, we hope that the reaction to the "Hot Canary" by the younger fry of the household will justify the time spent in building this novel project.

PARTS LIST

1 AC128 transistor or similar
1 AC125 or AC126 transistor or similar
1 Eagle TD3 transistor driver transformer LT44
1 Eagle TO3 transistor output transformer LT700
1 8 ohm 2" or 3" loudspeaker to suit case
1 Push-on / push-off or slider switch
1 9 volt battery (type 216 or similar) and connector to suit.

CAPACITORS

2 50uF/12 volt electrolytic capacitor
1 2uF/12 volt electrolytic capacitor
1 0.018uF polyester capacitor
1 0.0033uF polyester capacitor

RESISTORS

2 100K resistor, 1/2 watt 5 %
1 22K resistor, 1/2 watt 5 %
1 10K resistor, 1/2 watt 5 %
1 1K resistor, 1/2 watt 5 %

ASSORTED ITEMS:

1 Imitation bird (see text)
1 Plastic case (see text)
1 Aluminium panel to suit above (see text)
1 Piece 0.15 in Veroboard (see layout drawing)
Nuts, bolts, wire, speaker grille cloth, steel wire, contact adhesive, plastic foliage etc.

A SIMPLE FLASHER FOR 240V LAMPS

HERE IS A SIMPLE DEVICE WHICH CAN BE USED IN A NUMBER OF SOMEWHAT FRIVOLOUS APPLICATIONS, SUCH AS MODULATING THE LIGHTS IN YOUR CHRISTMAS DECORATIONS, OR FOR "DYNAMIC" MOOD LIGHTING AT A PARTY. IT COULD ALSO BE USED AS A PSEUDO-STROBE LIGHT IN A DISCOTHEQUE, WITH THE ADDED FEATURE THAT THE LIGHTS MAY BE ANY COLOUR YOU WISH.

While the device descrived in this article may be something of a gimmick the mode of operation is interesting and we are presenting two different circuits which do the same job. The project has the added feature that, if you have built the device and then decide that although it works it is, indeed, frivolous, you can use the parts to build something really practical, such as a light dimmer.

By way of explanation, the device does not flash incandescent lamps in the same way that strobe lamps are flashed, for in any case, the thermal inertia of a normal incandescent lamp filament makes it physically impossible to obtain a short, sharp flash. Rather, the power applied to the lamps is varied in what is basically a smooth, sinusoidal fashion so that, instead of flashing, the lamps can be said to be smoothly modulated at a steady rate.

Still, when rhythmically varied at a rate of, say, 5 to 10Hz, the lamps appear to flash as they are rapidly varied throughout their full range of brilliance. And when you are controlling thus the maximum allowable load of 500 watts of coloured light, it makes for what the younger generation call a "mind-blowing" experience.

Basically, the mode of operation is as follows. The heart of the unit is a Triac (bi-directional thyistor) which is phase-controlled. That is to say, the amount of power fed to the lamp is controlled by varying the point in each AC half-cycle at which the Triac is triggered into the conducting

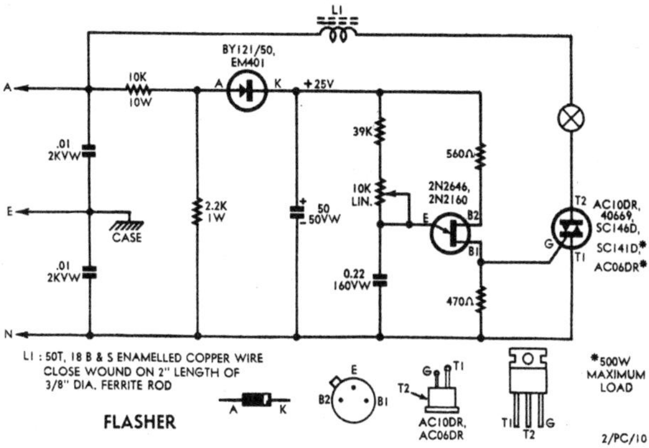

Figure 2, above, shows the recommended circuit using a UJT for triggering Figure 3, right, shows the Diac version.

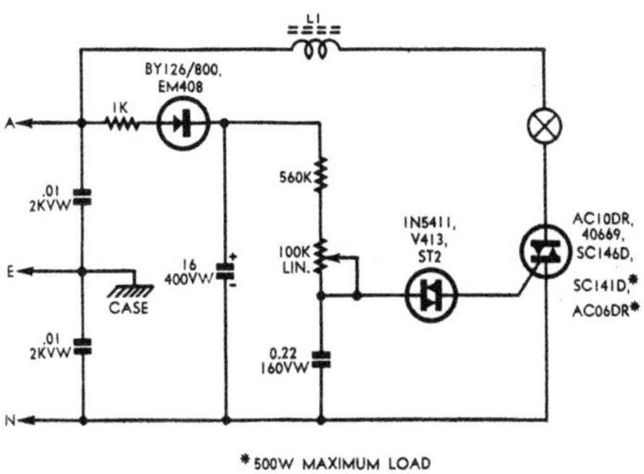

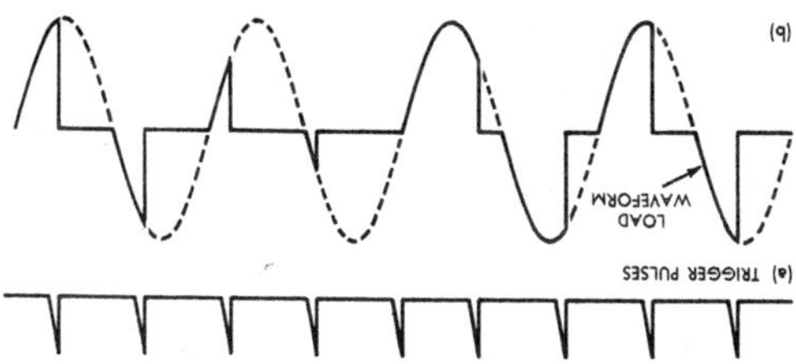

Figure 1, above, shows how flashing is produced by unsynchronised phase control.

state. If the Triac is triggered late in each AC half cycle the load power is low, while if triggering occurs early in each half cycle the load power is high.

To achieve a constantly varying level of load power we must have some source of trigger pulses whose phase, relative to the AC mains supply, is rhythmically varying. This is done by using a relaxation oscillator which generates trigger pulses at a rate of around, but not exactly, 100Hz. These trigger the Triac and so interact with twice the mains frequency (100Hz) to pulse the lamps at a rate equal to the difference between the relaxation oscillator frequency and twice the mains frequency.

The waveforms in figure 1 illustrate how the varying power level occurs. In effect, the lamps give a visual indication of the "beat" between the twice-mains and oscillator frequencies.

The first flasher circuit, figure 2, features a conventional relaxation oscillator using a 2N2646 unijunction transistor. This has its DC supply derived directly from the mains via a "half-wave" rectifier and 50uF electrolytic capacitor. The AC voltage is applied to the diode via a voltage divider consisting of a 10K and 2.2K resistor. Since a voltage divider was required at some point in the circuit we decided to place it before the diode as this enables a low PIV rated diode to be used and initial switch-on currents through the diode are kept low.

The effective DC supply for the UJT is approximately 25 volts, though this will depend mainly on the level of the mains voltage.

For those not familar with the operation of UJT relaxation oscillators, the following explanation may clarify matters. With reference to figure 2, the 0.22uF capacitor initially charges up towards the supply voltage via the 33K resistor and 10K potentiometer (connected as a variable resistor). The capacitor charges exponentially, eventually reaching a certain voltage between the potentials at base-1 and base-2 of the UJT, at which point the emitter/base-1 junction of the latter suddenly becomes a low negative resistance.

The capacitor is thus discharged via the emitter base-1 junction and a high energy current pulse results. Once the capacitor is discharged the emitter/base-1 junction again reverts to a high value of resistance so that the capacitor can be recharged. This cycle of events is repeated, ad infinitum. The result is a sawtooth waveform at the emitter of the UJT and a train of high energy, positive pulses at base-1 which are used to trigger the triac.

The 560 ohm resistor in the base-2 circuit of the UJT serves to improve the temperature stability of the oscillator.

During development of the UJT/Triac control circuit just described, it occurred to us that the same effect could be achieved with a simpler circuit using a Diac or symmetrical breakover diode. Normally, the Diac is used to phase-control triacs in light dimmers and similar applications.

Analysis of typical light dimmer circuits shows that, in effect, the Diac is operated as a simple relaxation oscillator which is syncronised to twice the mains frequency. What we need to achieve the flasher effect, is a relaxation oscillator operating at around 100Hz but not synchronised with the mains. This is easily achieved by supplying the Diac relaxation oscillator with filtered DC, as shown in figure 3.

The mains supply is half-wave rectified to give a DC supply of around 340 volts. The 1K resistor protects the diode against surge current at initial switch on. The 0.22uF capacitor is charged via the 560K resistor and 100K potentiometer until the voltage across it rises to around 30 volts, at which point the Diac breaks down and delivers a pulse to the gate of the Triac. Once discharge is complete the Diac reverts to its normal high resistance state so that the capacitor recharged. Thus the Diac circuit performs virtually the same function as the UJT circuit described above.

The only drawback of the Diac/Triac circuit just described is that is frequency stability is not as good as the former circuit. This means that having been set for a particular flash rate, it will drift about markedly over a short period of time. As a consequence it cannot be set to flash in time to the beat of the music, which the user may wish to do if it is to be used in a discotheque or at a party. Some draft is also noticeable in the UJT circuit, but is not nearly as serious.

We should perhaps point out that for both circuits there will be some setting of the rate control at which the relaxation oscillator will tend to lock with the mains, so that the lamps stay at a fixed level of brightness or may even go out altogether. This is quite normal but should not be any great disadvantage.

Of the Triacs specified, all but two may be used with incandescent lamp loads up to 1000 watts. The SC141D and AC06DR should not be used with loads in axcess of 500 watts. If sustained operation at ambient temperatures above 100 degrees Fahrenheit is envisaged, the circuit should be housed in a case with improved ventilation, so that the case temperature of the Triac does not rise above 90 degrees Centigrade.

While all the Triacs specified have adequate surge ratings to copy with their maximum specified loads, the individual incandescent lamps should not have ratings in excess of 100 watts. This is because the surge which occurs when a high power mains-voltage lamp burns out can exceed 100 amps before the internal fuse in the lamp blows, and this could easily ruin the Triac.

As with other Triac circuits employing phas control, the circuits presented tend to radiate electrical interference due to the extremely rapid switch-on times of the Triac. The suppressor inductor L1 and the two

.01uF capacitors bypassing the mains help keep the interference to a minimum. However, it is important that the unit should be built in an earthed metal case, as the inductor is an efficient radiator of the interference and should thus be shielded.

Incidentally, neither of the circuits presented is suitable for use with fluorescent lamps.

While both circuits were tried in the laboratory, we built up the unit featuring the UJT oscillator in prototype form to give readers a few hints on construction. The unit is housed in a sturdy diecast aluminium box measuring 4 5/8" x $3\frac{3}{4}$" x 2". Not only are they rugged but they automatically improve the effectiveness of the interference suppression components, provided of course that they are properly earthed.

The Triac is mounted on a small heatsink measuring $2\frac{1}{2}$"x$1\frac{1}{2}$", made from 18 gauge aluminium. Since the case of the Triac is at mains potential, the plate must be isolated from the earthed metal case by two insulating pillars. The AC10DR and AC06DR are metal cased units which are supplied with a mounting clamp, while the other Triacs specified are in the so-called "TO-66 plastic" encapsulation.

Some readers may be concerned at the fact that the full mains voltage is applied between the terminals of the potentiometer and the earthed case and potentiometer case, in both circuits. However, the manufacturers specifications indicate that a maximum of 1000V AC may be applied between terminals and case, so there is no real cause for concern.

The interference suppression inductor L1 is not available commercially but is quite easily made. Start by winding a layer or two of thin insulation tape on a 2" length of $\frac{3}{4}$" diameter ferrite rod. If a full length ferrite rod has been purchased it can be cut by filing a nick right around the circumference of the rod and then snapping it as it it were of glass - do not try to saw the rod. Close wind 50 turns of 18 s.w.g. enamelled wire over the insulation tape. Then wind insulation tape tightly over the rod in a couple of layers. The last step is important - if it is not wound tightly the inductor will make a buzzing noise due to currents being switched by the Triac.

Most of the circuitry is mounted on a suitable length of miniature tagboard. Layout is not critical but good wiring practice should be followed. The 10K resistor becomes quite warm and should not be in contact with the inductor L1 but rather should contact the bottom of the case, to improve heat dissipation.

The output to the lamps is connected via a normal 3-pin mains socket mounted on one end of the case. The wires to it should be passed through a hole of sufficient clearance to avoid any chafing - a 3/8" diameter hole would be suitable. The mains input cord should be passed through a grommetted hole in the case and anchored with a mains cord clamp. The earth wire should connect to the case.

And finally, here is a warning, it is generally recommended that persons subject to epileptic fits or migraine headaches should not watch intense pulsating light displays, as such conditions may be seriously aggravated thereby. Hence, devices such as those described in the foregoing should be used with some caution.

INSTALLING A BURGLAR ALARM

THIS ARTICLE WAS PROMPTED PARTLY BY A NUMBER OF REQUESTS WE HAVE HAD FROM READERS AND PARTLY BY THE ATTENTION DRAWN TO PROPERTY PROTECTION IN GENERAL BY RECENT POLICE STATEMENTS.

In a new house in a new suburb I had to assess the situation. At first there was little inclination to install a complete system because there seemed to be so many other things to do.

More recently, the situation has changed. What was once an "outer suburb" has now been overtaken by the suburban sprawl. With it came the realisation that the risk was now just as great as it had once been in the "inner suburbs". Any lingering doubts on this score were quickly dispelled by reports of court proceedings in the "local rag". Not only could it happen here; it was happening.

I made up my mind to get in first.

A new installation, in a house that was owned rather than rented, provided an excellent opportunity to apply some of the lessons learned previously. Broadly, these involved the following factors, in approximate order of importance. (1) Reliability of door and window contacts. (2) Battery drain. (3) Method of legitimate entry. (4) Concealment of wiring and contacts.

Before dealing with these subjects in detail, let us first consider the general concept of an alarm system and typical circuitry.

Most domestic alarms work on the principle of providing a very loud alarm usually outside the house, though well hidden, which will be clearly audible at least to neighbours on either side and preferably further away then this. Whether such neighbours are willing or able to summon assistance when the alarm goes off is really of little importance, because no prospective burglar will wait around to find out.

Most of them will cover the first hundred yards in even time, fences and other minor obstacles notwithstanding.

An alternative arrangement is one where it is desired to provide protection for a building remote from the main residence, such as a lock-up shop, garage, "ham shack" or workshop. In this case, it is preferable to provide a small bell, or even a buzzer, which will be audible only within the residence. In these circumstances an intruder need not know that his presence has been detected until a man in blue appears at his side and begins to ask certain awkward questions. Police officers to whom I have spoken on the matter agree that such systems result in a high percentage of catches.

It is usual with such systems to provide a large outdoor bell which can be switched into circuit when both premises are unattended, thus providing protection by the "scare" method.

Many commercial systems provide what might be termed the ultimate in protection by connecting each building to be protected to a central switchboard manned continuously by officers of the company providing the service. Connection from the building to the switchboard is by means of telephone lines provided by the Post Office and the occupier of the premises pays an annual rental for the entire service. Schemes like this, of course, are outside the scope of the amateur, but are mentioned as a matter of interest.

Comping back to more practical considerations for a domestic circuit we have to consider first whether to use a closed-circuit or open-circuit system. There are points in favour of both systems, although the closed circuit system is generally preferred.

The open-circuit system is one where the alarm bell is connected directly to the various door and window contacts so that the closing of any set of contacts, by opening the door or window, will cause the alarm to ring. Various schemes are used to ensure that the bell will keep ringing, even if the circuit is opened again immediately. The disadvantages of this scheme may be summarised as follows. Vital wiring of the system must be taken to each possible entry point, thus creating the risk of the system being rendered inoperative by a would-be intruder cutting the wires or tampering with the contacts.

A second point is the possible failure of a particular set of contacts to close the circuit properly. To ensure that the system is always working it is necessary to test all the contacts at fairly frequent intervals. Many non-technical owners do not fully appreciate this and are content to test the circuit by opening an odd door or window and being satisfied if the alarm operates.

On the credit side it must be conceded that such systems are less prone to false alarms than other systems, as well as being lighter on battery current. In fact, the batteries should have a life almost equal to their shelf life, since they will have only a brief drain imposed on them for testing at well-speced intervals.

The closed-circuit system operates on the principle of using a circuit through the window and door contacts to hold in a relay. A pair of "normally closed" contacts on the relay are thus held open and, since they form part of the main bell circuit, the bell is prevented from ringing. Opening a door or window will open the relay circuit, allow the relay to drop out, and close the main bell circuit. As with the open-circuit system the bell will continue to ring until a local circuit is interrupted. Obviously there is little chance of the contacts failing to operate, while a cut wire will also operate the alarm.

In all fairness it must be admitted that the closed-circuit system also has disadvantages. The natural tendency is to a false alarm if the contacts are faulty and too many of these can nullify the whole purpose of the scheme. Since there is a continuous drain on the batteries while the system is set up they will have a much shorter life, though most people still consider that the few cents a week running cost is cheap insurance.

All things considered, and assuming that the weak points of the system are recognised and allowed for in designing the installation, the closed-circuit system is to be preferred.

The accompanying circuit will give some idea of the practical application of the principles just explained. It will be seen that the battery, door and window contacts, the relay coil, and a set of contacts on the relay all form a series circuit. The purpose of the relay contacts is to provide the continuous-ringing feature and their operation is quite easy to follow. Assuming for the moment that the relay is being held in, it can be seen that the current through the coil must flow through the relay's own contacts in order to hold it in. In other words, the contacts are what are known as "self-holding" contacts. It is easy to appreciate that such a relay cannot pull in again once it has dropped out because it will have opened its own coil circuit. Thus, once the alarm is started, it cannot be stopped by simply restoring the contact circuit.

Obviously, some method must be provided to cause the relay to pull in whenever the system is to be set. One way would be to simply push the relay in by hand, but this has the disadvantage that the relay may not always be conveniently located, plus the far more important problem of protecting the relay contacts. Because of the inductive nature of the circuit, due to the relay coil, considerable sparking can occur when the circuit is broken. Theoretically, this should not be a problem when the relay is being pushed in but, in practice, it is possible for the current to vary both up and down at the moment of contact, particularly if the person concerned does not appreciate the need to make a quick and positive movement.

A far better arrangement is to connect a push button across the holding contacts so that they are momentarily short circuited as the button is pressed. This causes the relay to pull in, holding in on its own con-

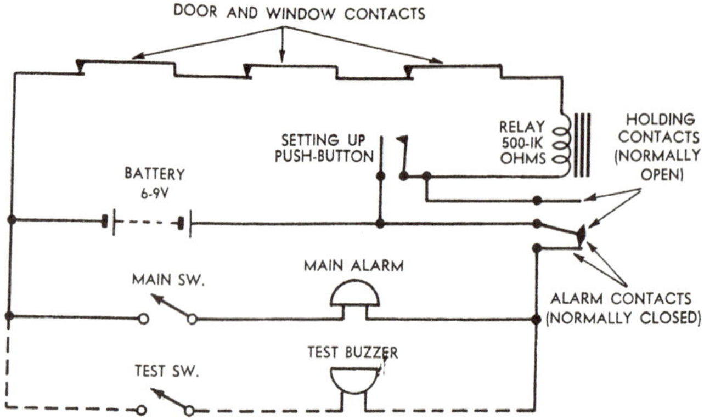

The basic circuit of an alarm system. The buzzer may be used to check that the relay is set, or a small globe may be substituted for it. Alternatively, the buzzer may be used for night time monitoring.

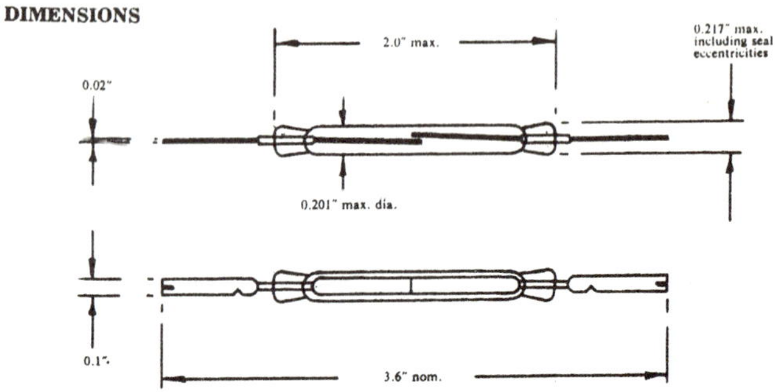

The heart of the system is the dry reed switch. The Hivac type XS7, illustrated here, has been specially designed for the low current conditions and prolonged closure times inherent in alarm systems.

cut the recess, and it will be found that the tenon is usually very close to the same size as the magnet. The magnet will not be obtrusive in this position since the tenon normally breaks the smooth surface anyway.

Next, the wiring. How this is carried out will depend a lot on the type of house, i.e. brick, weatherboard, fibro, etc. but some factors are common. For example, it is worthwile going to some trouble to conceal the wiring, usually by getting it into the wall cavity from each point, and then either into the ceiling or under the floor for routing to the control panel. With a little care it is possible to install a system which will display no clues as to it existence.

Just how important this is is the subject of some debate. There are those who feel that there is merit in advertising the existence of an alarm, on the basis that this will discourage a potential burglar. On the other hand, the writer feels that the element of surprise is the householders strongest weapon. A burglar who knows an alarm system exists is half way towards finding a way to circumvent it, assuming he feels that this is worthwhile.

And, particularly where shops and stores are concerned, the value of the contents is often quite sufficient to tempt a burglar to "have a go" even if he has to spend a lot of time sizing up the situation before he strikes. Any part of an alarm system which is visible not only warns of its existence, but may easily give a clue as to how the system works and, therefore, how best to beat it. The less that is visible therefore, the better. The following points are based on the writer's experience, but obviously cannot cover all situations. Where strange ones are encountered, the individual can probably work out a solution for himself.

Where it is planned to locate a switch in the top of a door frame, it is usually worthwhile to remove the outside architrave on the lock side of the frame. This will give access to both the side (directly) and the top (indirectly). If the switch is located close to this top corner of the frame - which is a good place anyway - then it is a simple matter to feed a "fish wire" into the space above the frame through a hole in the channel, coax it out the end where the architrave is removed, and use it to draw the circuit wire back into the channel. If the wire is to go under the house it may be run down the side of the frame and through the floor.

If it is to go into the ceiling it should still be possible to gain access to the wall cavity from the corner of the door frame, but it may be easier to feed a fish wire down from the ceiling, "hook it" from the access point at the frame corner, and then use it to draw the circuit wire back up into the ceiling.

Removing an architrave normally is not difficult. Painted architraves (such as the outside ones) should be selected in preference to stained or polished ones, since any minor damage is easily repaired with putty and paint. If old nails have to be withdrawn from timber before it is replaced, less damage will be done if the heads are pulled right through the timber from the back, rather than driven out the front.

In the case of box frame windows the best approach seems to be to get the wiring into the weight cavity, on the most convenient side in the case of a single window, or the centre in the case of a double window. From here it is brought out the front of the cavity through a hole drilled at sill level, then through the sill via another hole. A certain amount of improvised "routing" with the drill, in the vicinity of the junction of these two holes, will allow the wire to be recessed. The holes may be filled with putty or epoxy.

Once under the sill there are a number of possible ways in which the wire may be routed. One of the easiest is to simply force it between the bottom of the sill and the moulding beneath it, prising these two pieces of wood apart slightly if necessary. This will allow the wire to be run to the end of the sill, or any intermediate point where another path is available. Where fibro cement is used a cover strip may be lifted and the wire run under this either up or down as required. Before embarking on such work, make sure that matching paint is available to repair any minor damage that may occur.

In the case of brick walls, it has been suggested that a channel, deep enough to accommodate the wire, could be scraped in the mortar, then sealed with a fresh coat of mortar. The success of such a scheme would depnd on how well the new mortar could be made to match the old, and the skill with which the job was finished. Weatherboard finishes are probably easiest to deal with, since any holes or channels which are needed can usually be filled and finished using normal woodworking materials and techniques.

The technique of "hooking" wire feed into a cavity is extremely useful, though it sometimes calls for a little patience. On an average however, one will usually hook the wire within a few minutes. The writer used a hook made from a length of soft iron packing case wire, about 16 gauge. A reasonable size hole should be drilled in to top or bottom wall plates, say 3/8" or $\frac{1}{2}$" using an auger bit. Make the hook as large as possible, consistent with the need to go through the hole, and feed as much wire as possible into the cavity from the other end. This can be scrap wire it necessary.

In fishing for the wire, let the hook wire assume some curvature so that it approaches the horizontal as it is fed in. Then, if the first few tentative tries are unsuccessful, bend the wire into a crude crank and rotate the hook so that it thrashes around inside the cavity. This usually brings results.

The most convenient wire to use for the contact circuit is ordinary bell wire, preferably in "twin" formation. The official description of the wire used by the writer is "1/.028 Twin Figure 8 Extra L.T. P.V.C. Insulated Bell Wire." It is available in 100yd rolls. Its small size makes it very easy to coax it into cracks between timber, fibro, etc.

In actually running the wires from each point to the control panel, one has a choice of two approaches : (1) Run a single wire from each point to the next, forming a single loop right round the building, or (2) bring each point - or group of points - back to the control point as an independent pair, the various pairs then being connected in series at the control point.

Arrangement (1) has the advantage of simplicity and that it uses a minimum of wire, but (2) has some advantages which may be worth considering. The main one is that it becomes possible to determine with a meter system, from the control point, just which entry point (or group) is open circuit in the event that the alarm is tripped, or will not set. The latter is the most useful asset. When setting up the system before leaving the house it is most disconcerting to find that it will not set, due to some window being not quite closed. Much time can be saved if one can determine even the room in which the defect is present.

It is even more disconcerting if the alarm raises one from a sound sleep in the middle of the night - assuming the worst should happen - and, having switched it off while still half asleep, one stands by the control panel wondering where the heck the thing has been tripped. This is particularly so where outbuildings form part of the system.

In any case, where outbuildings are involved, it is desirable to keep these circuits separate beyond the control panel. Then, by suitable switching, it is possible to set the alarm for this part of the system only, providing protection for the remote buildings at any time that the main building is occupied.

The writer has not provided such refinements yet, but has wired the system for them on, roughly, a room by room basis. The indicator will probably have to be meter movement (one of the cheaper variety), capable of being switched across each circuit. This may double as a voltmeter for battery checking.

As intimated earlier, the whole success of an alarm system depends on its ability to make plenty of noise. For this reason a really good bell is essential and the inexpensive types normally used for door bells are quite useless. What is required is a large bell, specially designed for this kind of work and having at least a 10" diameter gong. It should also be a "weather clad" type. Such bells are normally available from electrical supply houses or from firms associated with commercial burglar alarm installations. They are not cheap, but are still a good investment, when one considers the likely loss involved in a single burglary.

They are available in a range of voltage ratings, from six to 240, the six volt types being normally used for domestic installations. However, many professional systems use a 6V bell with a 9V battery, to increase the sound from the bell. Such bells seem well able to cope with this ove-load.

Alternative forms of alarm which might be considered are motor car horns and relatively high powered audio oscillators operating into a large loudspeaker. An advantage of either of these systems is that they lend themselves to concealment behind a wall with only a small grille type opening for the sound. They may also be cheaper than some bells.

A disadvantage of the car horn is that it draws a lot of current - more than could be supplied by dry batteries or handled by the normal type of relay. An accumulator and an extra relay, of the type used for this function in cars, would be essential. Also, the sound made by either the car horn or the oscillator, may not be as readily recognised as a burglar alarm, as would the bell.

Whatever kind of alarm is chosen it must be located where it can be heard but not seen or be readily accessible by anyone disposed to temper with it. Some commercial systems provide two bells, so connected into the system that any attempt to tamper with one will trip the relay and set the other ringing. This may be more elaborate than is justified in many cases, but may be considered where it is difficult to conceal the bell adequately.

In one installation which the writer has seen, the bell was located under an iron roof. Most of the loss of sound which such a position would normally create was overcome by arranging that a particular part of the gong rested against the iron. The position on the circumference of the gong was carefully selected so that the presence of the iron did not dampen the vibration of the gong, but rather conveyed the vibrations to the iron. As a result, the whole roof resounds to the voise of the bell.

The wiring to the alarm should introduce a minimum of resistance in order that the best use be made of the available battery voltage. Twin nylex sheathed power cable (1/.064) is eminently suitable and, if fitted with an earth wire as well, this can be paralleled with one of the other conductors to further reduce resistance. This is particularly desirable if a long run is involved.

Battery operation is virtually essential for any alarm system. Mains operation leaves the system vulnerable to anyone who cares to turn off the main switch, which is normally readily accessible. Also, in the event of a mains failure, even for a few seconds, they relay will drop out and set off the alarm when the power is restored. The only question which has to be decided is which type of battery should be employed.

Commercial systems prefer dry batteries for the fairly obvious reason that, apart from the need to renew them at regular intervals, they requre no maintenance. A typical battery would consist of from four to six large cells. Used in a typical closed circuit system where it is set every night and for several hours a week during the day, such a set of cells should last from nine to 12 months. More recently, circuits employing solid state devices in place of, or in addition to, the relay have been developed, and these can effect a marked reduction in current drain. So far, the writer has not progressed further than scribbling a circuit on paper, but the possibilities seem very promising. Such a circuit may well swing the balance completely in favour of dry batteries.

The alternative approach is to use a small accumulator floated across a trickle charger adjusted to keep it near full charge at all times. While cheaper to run than dry batteries, it will probably cost more initially. In addition, all such systems require some maintenance, rendering them unsuitable for non-technical users.

A large accumulator is not required, since even a small unit would operate a typical alarm bell for several hours. A motor cycle battery would be a good proposition where six volts was considered adequate for the bell or other alarm. If a higher voltage is desirable, then four single cells, giving a nominal eight volts - or a little more if on charge - would be the best approach.

In any such system the charge rate will need to be adjusted to just offset the battery losses and the drain from regular use. It should also be capable of charging at a faster rate in the event that the alarm is operated and the battery well discharged. The best way to adjust the float rate is to take frequent gravity readings, say once a week, until a minimum rate is found which will allow the gravity to be maintained. After this less frequent readings should be sufficient.

The relay for this circuit is a fairly standard type which should be available from anumber of sources. The contact requirement is for one "normally open" set (for the holding circuit) and a "normally closed" set (for the bell circuit). These may be separate sets or combined in what is called a "change over set". The circuit shows the latter arrangement in use. The bell circuit will need to handle at least one amp in most cases, and this may be rather high for the small silver or platinum contacts normally fitted. Heavy duty silver contacts rated at about 3A would be better.

Although the bell circuit is inductive, the contacts will normally only ever have to close this circuit, which is much less exacting than opening it. The opening function will normally be performed by the main bell switch. The holding contacts have an even easier task; only a few milliamps of current and no requirement to either make or break a circuit.

In the interests of reliability the relay contacts, and particularly the holding contacts, should be protected from dust. A modern fully enclosed relay would be ideal but, failing this, mount the relay with the contacts vertical, so that there is less chance of dust settling on them. Consideration was given to using a suitably rated dry reed switch in this application, but investigation showed that the efficiency is relatively poor and that the battery drain would therefore be excessive. However they may be suitable for use in conjunction with a thyristor, as already mentioned. The writer hopes to investigate this circuit as soon as time permits.

The resistance of the relay coils should be as high as possible in order to keep battery drain to a minimum, but no so high that its holding qualities are endangered. Relays vary in their efficiency, but, as a general rule, a 500 ohm coil should suit a 6V system and a 1000 ohm coil a 9V system.

Some means must be provided for the occupant to leave and re-enter the building without setting off the alarm and without the need to create a vulnerable point in the form of an unwired door. Commercial installations usually provide a multi-pin plug and socket is used, such as the octal type, quite a complex system of cross connections is possible in both halves making it virtually impossible for anyone to pick such a combination.

The advantage of this scheme is that, even if another person sees it used, they are little the wiser as to how it operates. Also, the presence of the plug in one's pocket is a constant reassurance that one has not forgotton to attend to this part of the alarm system. A disadvantage is that the plug can be lost or mislaid while out, resulting in the need to trip the alarm, perhaps late at night, in order to re-enter the house.

The switch system avoids these problems, but has one of its own which may worry some people. If the switch is left set after the house is re-entered, an astute person may learn the code by simple observing the position of the knob, relative to any marks on it, orby turning it back to the stop. Such a situation is not very likely, but may be overcome by using a switch requiring a two-figure combination.

The writer made such a switch, more for the satisfaction of doing it than anything else, using standard multi-position switch parts. This has proved completely satisfactory in service. It is really two complete switch movements, consisting of clicker plate and 11-position wafer, connected in tandem. The secret is the coupler between the two sections which introduces a deliberate backlash of several positions. Thus it is possible to set the rear switch to one position by moving the knob, say, clockwise, then the front switch to another position by moving it anti-clockwise. The two positions selected are connected in series, and the whole combination in parallel across the door contacts.

The first attempt at such a switch ignored the need for a clicker plate in the rear section, but this proved unsatisfactory and was never used. The main problem was a tendency for the rear section to "follow" the movement of the front section due to unavoidable friction between the shaft and the bush. This could have led to false alarms.

The coupling between the two sections was made from a pair of standard $\frac{1}{4}$" to $\frac{1}{4}$" brass couplings, soldered together. A little more than half of one section was cut away, to give a backlash between the two movements equal to about five positions. This arrangement can be varied as desired. Some bushes may be long enough to eliminate the need to join two together gether. The pin through the shaft of the front section is a section of shank from a broken drill, the same size drill being used to make the hole for it. It should be a drive fit.

Such a switch may be mounted so that the knob is located on the architrave, or on the door itself. However, the latter arrangement will call for flexible leads between the door and the frame. The switch itself may be located behind the architrave, or on the inside architrave with a long shaft to the outside knob. The latter arrangement is probably safest, but calls for a more compact version of the switch than the rough version shown in the photograph. It may then be housed in a neat metal box, and may also have an additional knob on the inside. This allows the system to be set up from inside on the way out.

Finally, having installed the system, make sure good use is made of it. Set it up every time the house is vacated, even if you only intend visiting the next door neighbour for a few minutes. It should also be set up at night, because an experienced pilferer can enter and ransack a house without disturbing even a light sleeper.

MORE ABOUT THE BURGLAR ALARM

The first point concerns the use of the alarm at night, while the premises are occupied, and the problem of providing an open window - or windows - for ventilation. Should that section of the alarm be disabled and if so how, or must the whole idea of using it in these circumstances be abandoned?

It seems to the writer than an open window at night is the very think that needs protection, since it is a favourite entry point for prowlers, and not only those bent on material gain. Therefore, rather that bypass this section of the system, every effort should be made to keep it in operation, in spite of problems it may present.

One suggestion is to fit an additional magnet into the side of the window to work in conjunction with a reed switch recessed into the outside of the weight cavity at a suitable height above the sill. This reed should be connected in parallel with the one in the sill. In this way, the window may be opened a set amount, chosen as sufficient for ventilation but in-sufficient for a person to enter without raising the window further.

It is a good idea to use one of the larger magnets (FM448) for this application, to provide a margin against side ways play of the window, and minor errors in opening, although the correct position for the latter should be marked in some unobtrusive way.

Possible objections to this idea are the fact that the window can be opened only a set amount, which may not always be optimum for the weather conditions, the fact that the window could not be closed in the event of a sudden change in the weather, without switching off the alarm, and the possible difficulty of fitting the magnet to the side of the window. Never theless, there may be many situations where these are not serious.

Another idea is to include the wire insect screen in the alarm system. If this uses a wooden or aluminium frame a magnet can easily be fitted to it to work in conjunction with an extra reed switch in the sill. As before, this should be wired in parallel with the window reed. Since most such screens are simply clipped in place it is almost certain that a would be intruder would simply unclip the frame so as to gain entry with the least possible noise. However, the possibility that the wire may be cut cannot be overlooked, particularly if it is plastic rather than metal.

Finally, a suggestion which has no electronic tie-up at all, but which might logically be used in conjunction with either of the above ideas. This is to fit one of a variety of locking devices which either lock the two halves of the window together in some selected relationship or locks them to the adjacent frame.

The general principle of the former category is to drill all the way through one window frame and part way through the other, then lock the two together by means of large wood screws, metal screws through tapped metal plates, or simply a length of plain metal rod. In the second category, the window proper is locked to the frame by means of ordinary tower bolts. This arrangement has the advantage that, by drilling several holes in the frame the window may be locked in any one of several positions.

A rather different problem which the writer encountered recently may prove of interest to readers. The back door of the house opened into the laundry, from which a second door led to the house proper. The owner wanted to be able, on occasions, to leave the outer door unlocked so that trusted trades people could leave goods if they housewife was out. On the other hand, he didn't want the outer door completely by-passed, since it would leave the laundry - and a valuable washing machine - vulnerable when the house was vacated during holiday periods.

While the problem could have been solved by means of suitable switching, it was felt that this should be avoided if possible, due to the possible confusion it might cause to inexperienced people, particularly the housewife. The solution was simply to wire both doors and connect them in parallel. Thus, when the outer door was to be left unlocked, the inner door was closed and locked, thereby by-passing the outer door and allowing it to be used. When the outer door was to be locked, the inner door was naturally left open.

Finally, from a friend came two suggestions for reducing current drain while the system is set. Both are based on the fact that a relay needs less current to hold it in than it does to pull it in.

Figure 1 shows the simplest arrangement, where a resistor is so connected that it is by-passed by the setting-up button, but comes in to circuit in series with the relay coil when the button is released. The resistor is made as large as possible consistant with reliable holding.

Figure 2 calls for a special relay, although this type has been available through disposals sources readily enough. It has two windings which, as shown, are of unequal value. However, even a pair of equal value windings would offer some advantage. Once again, the connections are such that the setting-up button excludes the extra circuitry until the relay has pulled in and the button is released.

Both these arrangements are worth considering and, in fact, I tried something very similar to figure 1 in the first installation I built, and which operated from dry cells for a time. The important requirement is summed up in the phase "reliable holding" since, if the idea is carried too far, there is serious risk of false alarms.

Of the two ideas, figure 2 is the more efficient, since there is no waste of power in a resistor which contributes nothing to the holding ability of the relay.

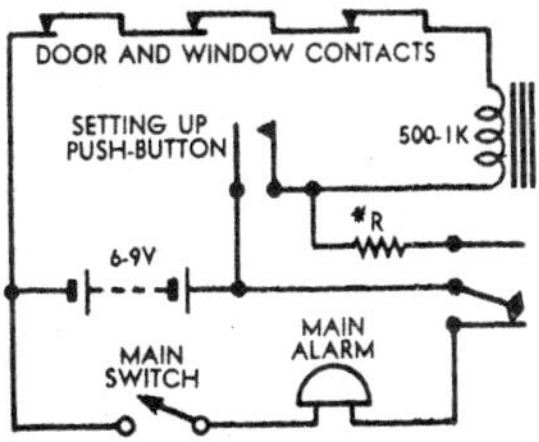

Figure 1. A simple method of reducing standing current. The resistor value must be such that the relay holds in reliably.

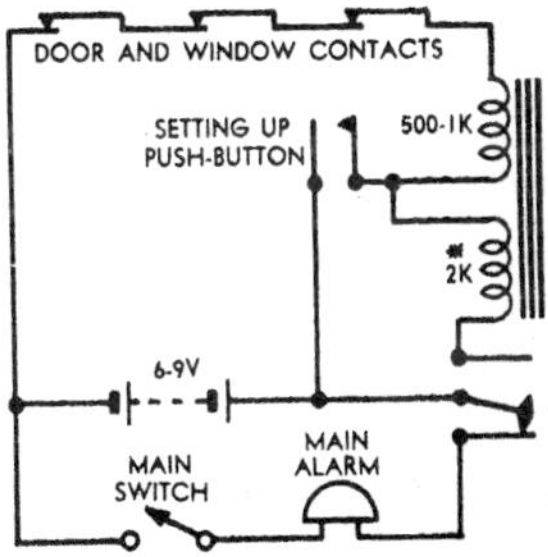

Figure 2. A better method using a relay with two windings. There is no power waste in an external resistor.

TRACING BURIED PIPES AND CABLES

IT WOULD BE TRUE TO SAY THAT IN AREAS RETICULATED FOR ELECTRICITY NEARLY ALL BURIED METAL PIPES CARRY CURRENT AT MAINS FREQUENCY, PARTICULARLY IF THE MULTIPLE EARTH NEUTRAL SYSTEM IS USED. THESE CURRENTS GIVE RISE TO ALTERNATING MAGNETIC FIELDS WHICH ENABLE THE PIPES AND CABLES TO BE TRACED IF A SUITABLE DETECTOR IS AVAILABLE.

The detector about to be described consists of three main parts, a pickup coil, an amplifier and a pair of headphones. The pick-up coil has an impedance, at 50 cps, which roughly suits the input imped-

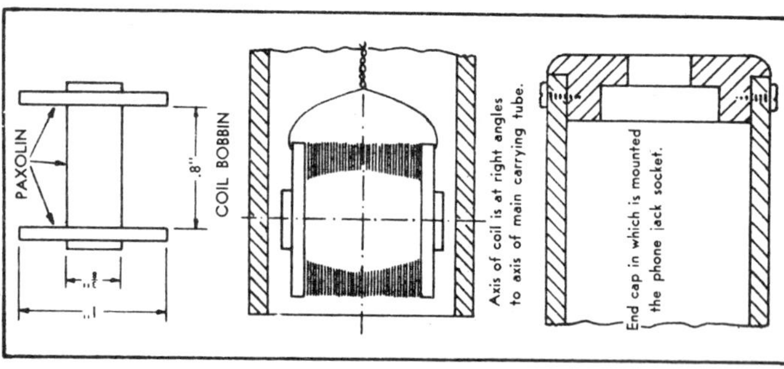

Detail sketches of the coil assembly, its position inside the lower end of the tube and the cap which seals the upper end of the tube.

ance of the amplifier and has a mu-metal core, which enables it to respond to very small magnetising forces. If mu-metal cannot be obtained, Stalloy or similar material may be used with some decrease in effectiveness.

The amplifier is designed to introduce distortion products into the 50-cycle waveform such that plenty of signal is produced in that part of the audio spectrum favoured by the human ear. A linear amplifier would be most unsatisfactory since neither the ear nor the headphones respond very well to the frequency of the mains. The amplifier has useful gain over a reasonable range of ambient temperatures.

The coil and amplifier are housed in a bakelised paper tube 30" long and 1 3/8" internal diameter. The wall thickness should be great enough to provide adequate mechanical strength and it has been found that 1/8" is satisfactory in this respect. The phone jack plugs into a jack socket mounted into the end cap of the carrying tube. Inserting the jack completes the battery circuit to the amplifier and thus removes the necessity for a separate switch.

This method of construction probably represents the optimum with regard to convenience in use but other approaches are possible. For example, the pick-up coil could be mounted at the end of a rod and the amplifier contained in a small box which would be carried in the coat pocket or suspended from the shoulder by means of a strap.

The pick-up coil is wound on a bobbin to the dimensions shown and the winding space filled with 38 s.w.g. or 40 s.w.g. enamelled wire. Three mu-metal laminations 1" x $\frac{1}{4}$" are inserted in the bobbin tube and serve as the core. It is interesting to note that three laminations provide about 95 per cent of the possible core gain so do not waste this useful material by filling up the bobbin tube.

After providing suitable anchorages for the coil leads, cast the coil in epoxy resin so as to form a solid cylinder which is a sliding fit in the main carrying tube, i.e. just under 1 3/8" outside diameter. This unit may now be cemented into one end of the carrying tube by means of an adhesive.

Arrange that the twisted coil leads are long enough to project about six inches beyond the remote end of the carrying tube so that they may be later connected to the amplifier.

The amplifier is built on a piece of Paxolin 1/16" thick, $6\frac{1}{2}$" long and just under 1 3/8" wide. Leave about $\frac{1}{2}$" of the Paxolin free at each end for cementing into the wooden supporting cylinders. One supporting cylinder is recessed to take the bakelised paper tube which forms the battery compartment. This tube has an internal diameter of 7/8" and a wall thickness of 1/16".

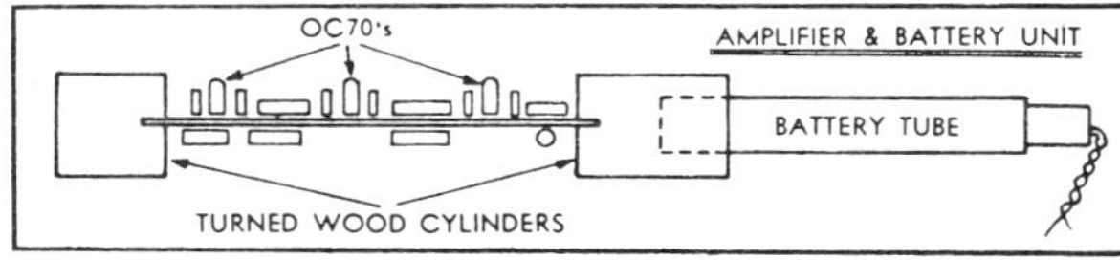

Shown below is the circuit for the transistorised pipe tracer. Unlike our "TV-AID," with which it bears some comparison, the amplifier has no gain control, since there is no need to prevent signal overload. The sketch above indicates broadly how our contributor constructed the amplifier on a piece of flat insulating strip, fixed between two cyclindrical wooden supports.

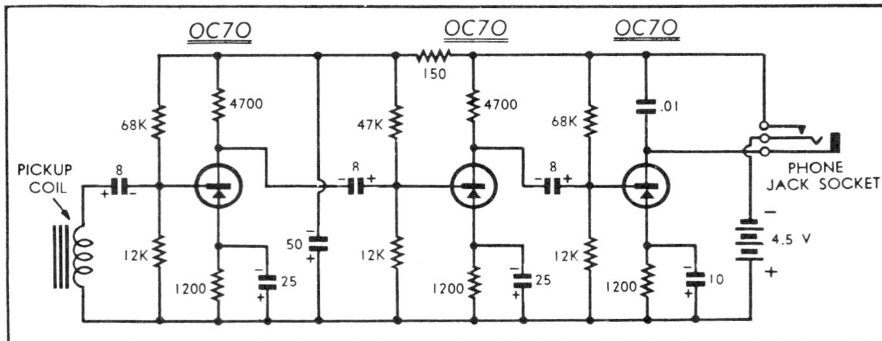

A coil spring provides the battery negative connection at the inner end and a wooden bayonet type plug serves to close the open end as well as providing a means for making the positive connection.

All resistors in the amplifier are $\frac{1}{4}$ watt and the capacitors are miniature electrolytics with the exception of the phone by-pass which is a disc ceramic.

The component leads provide all the wiring necessary in the amplifier itself with the exception of one tinned copper wire which runs near and parallel to one edge of the Paxolin strip and forms the positive common busbar.

The end cap carrying the jack socket and the amplifier assembly are secured in position in the main carrying tube by means of small wood screws or "self tappers".

This detector unit may be used to trace all types of underground metal pipes or cables provided these are carrying a small amount of alternating current or DC having rectifier or commutator ripple.

On farms, gold course, etc., which are remote from an electrical reticulation system, underground pipes will not carry AC and it will be necessary to inject a signal so that they may be traced. The simplest signal generator is possibly a battery plus bell or buzzer. Two leads should be taken from across the bell contacts; one lead is connected to the pipe system at any point where it is exposed and the other to an earth spike driven into the ground about 15 feet to one side of the pipe run.

An audio signal generator of the type used in radio work also makes an excellent signal source. The frequency should be set to about 1,000 c.p.s. With a suitable injected signal, pipes may be traced up to 500 yards from the signal source.

The detector may also be used for tracing cables, conduit, etc., in the walls and floors of buildings. If it should be necessary to provide tracer current, use one of the methods outlines in the preceding paragraph.

Short circuits in cables can be located by tracing the signal until it fades out at the fault. For this application the signal path should be along the faulty conductor, through the fault and back to the generator via the other faulty conductor.

AN ELECTRONIC DICE

THE GADGET DESCRIBED IS IN A SENSE A LOGICAL FOLLOW-ON FROM THE "ELECTRONIC DECISION MAKER" DESCRIBED IN THIS BOOK : IT IS AN "ELECTRONIC DIE" OR AS MOST PEOPLE WOULD SAY, AND "ELECTRONIC DICE". QUITE APART FROM ITS INTEREST AS A NOVELTY PROJECT, IT CAN SERVE AS AN INTRIGUING TUTORIAL EXERCISE FOR THOSE WHO WANT TO FAMILIARISE THEMSELVES WITH DIGITAL CIRCUITS AND INTEGRATED LOGIC MODULES.

Whereas the Electronic Decision Maker was virtually an artificial "coin", producing simply a "tails - heads" or "yes - no" indication, the gadget described in this article produces a "one out of six" indication and therefore corresponds to an artificial "dice".

In order to produce a device capable of performing this function we have found it necessary to employ a circuit a good deal more complex than the simple neon tube multivibrator used in the earlier project. However, the use of modern itegrated microcircuits ("IC's") has kept costs to quite a modest figure, while constructionof the project itself can prove of considerable value in terms of a tutorial exercise in digital circuitry and logic principles.

We doubt that many readers want to construct the unit purely to possess it as an electronic gimmick. Readers merely wanting a dice should go out and buy one! There is, also, the difficulty of convincing the layman that electronic devices of this kind are not biased.

THEORY OF OPERATION

Basically the unit consists of three so-called "J-K flip-flops" connected in a twisted ring counter configuration and fed with pulses at 50Hz. The output of the flip-flops is used to drive six lamps.

A flip-flop is a variety of bistable circuit element; as the name suggests it is a circuit configuration which has two stable operating states. External signals can be used to switch such an element from one state to the other and back again, as desired, the element remaining between the application of signals in the operating state which resulted from the preceding switch-over. Thus the element has "memory", or a capacity to store a "yes-or-no" item of information.

Quite a few devices and circuit configurations have two stable states. The tunnel diode may be operated in this fashion, as may neon lamps and PNPN devices. Similarly, toroidal ferrite cores may be arranged to be stably magnetised in either direction. However, the most common bistable elements are valve or transistor "flip-flops". Of these, the J-K flip-flop has the largest number of potential applications by virtue of its greater flexibility and fully predictable behaviour.

As the flip-flops in our device we used three micrologic ICs from Fairchild, type Ful923 J-K flip-flop. These are supplied from a 3.6V source and are guaranteed to operate reliably at frequencies up to 2MHz.

The circuitry inside a micrologic J-K flip-flop is shown in figure 1, together with its pin connections and the corresponding logic diagram symbol. As may be seen the outputs of the flip-flop are designated as Y and Y-complement, where the latter output is the logical opposite or "complement" of Y. In actual fact this means that when one output is at a "high" potential (nearly +3.6V) the other is at a near-zero or "low" potential, and vice-versa.

In logic circuitry "high" and "low" potentials are often equated to logical "truth" (1) and "falsity" (0). Just which of the two potential levels is regarded as equivalent to which logical value depends upon the "logic polarity convention" adopted.

The J and K terminals of the flip-flop are gating inputs, whose potentials determine the behaviour of the flip-flop when an input pulse is applied to the "toggle" or triggering pulse input T. Both J and K can either be at the high potential level or at the low potential level, where the potential levels may be interpreted as logical values as shown in figure 1.

Astute readers will have noticed that the logic conventions for the micrologic slip-flop differs from that of the discrete component flip-flops featured in the recent series on Logic and Counting. The difference lies in the fact that the logic convention for the J and K gating inputs of the micrologic flip-flops is identical with that of the outputs, whereas that for the J and K inputs of the discrete component flip-flops was the reverse of that of the outputs. This reversal of polarities occurred because of the difficulty of designing a discrete component flip-flop that would have consistent logic convention while keeping the number of components to a minimum. Note that the truth tables still apply, in spite of differences in logic convention.

The behaviour of a J-K flip-flop can be predicted for any set of conditions by compiling a so-called "truth table" as set out below. The subscript "n" is used to indicate the outputs of the flip-flop before the application of the "toggle" pulse, and the subscript "N + 1" to indicate the outputs after the application of the "toggle" pulse.

See Fig. 1A

An explanation of the mechanism of flip-flops is beyond the scope of this book and, as can be imagined, the mechanism of operation of the micrologic devices we are using in this project is considerably more complex than that of comparable flip-flops using discrete componentry.

J	K	Y_n	$\bar{Y}_n$	Y_{n+1}	$\bar{Y}_{n+1}$
0	0	1	0	1 } Y_n	0 } $\bar{Y}_n$
0	0	0	1	0	1
1	0	1	0	1	0
1	0	0	1	1	0
0	1	1	0	0	1
0	1	0	1	0	1
1	1	1	0	0 } $\bar{Y}_n$	1 } Y_n
1	1	0	1	1	0

Fig. 1A
Truth Table

Input Pulse	FFA	FFB	FFC
0	$\bar{A}$	$\bar{B}$	$\bar{C}$
1	A	$\bar{B}$	$\bar{C}$
2	A	B	$\bar{C}$
3	A	B	C
4	$\bar{A}$	B	C
5	$\bar{A}$	$\bar{B}$	C
6	$\bar{A}$	$\bar{B}$	$\bar{C}$

Fig. 1B

1	$\bar{A}$	$\bar{C}$
2	A	$\bar{B}$
3	B	$\bar{C}$
4	A	C
5	$\bar{A}$	B
6	$\bar{B}$	C

Fig. 1C

As we have already said, the Electronic Dice involves three J-K flip-flops connected in a "twisted ring counter." A twisted ring counter is a special type of ring counter which, in turn, is a development of the "shift register". At this point, the novice may well fee confused but please bear with us.

A shift register is a series of bi-stable elements interconnected in such a way that, when pulsed simultaneously by the input signal, they each behave in a manner determined, not by their own previous state, but by that of the preceding element. The connections are made in such a way (Y of the first flip-flop connected to J of the second flip-flop and Y-complementary to K) that, when the register is activated by the signal pulses, each element switches to the state previously occupied by the preceding element. Thus a series of input ("clock") pulses will progressively shift a pattern of values along the register.

A ring counter is essentially a shift register in which the two ends have been connected to form a closed loop. A pattern of logical values is circulated around the "ring" by the input pulses, so that it is possible to note the number of pulses received by observing the position of the logical value pattern.

The essential difference between the normal ring counter and one that is "twisted" is that the end-to-end connections are made so that there is a logical twist or inversion.

The twisted ring counter is more suitable for counting than the normal counter, because a twisted ring counter does not have to be supplied with an initial pattern to circulate. If a normal ring counter is reset to zero, the flip-flops will not switch over at the application of the toggle pulse because they are in a uniform state; consequently no pattern will be circulated. The twisted ring counter can count from a reset or all-zero condition, by virtue of the "logical twist" in the end-to-end connection.

Reference to the schematic of the Dice should clarify the concept of the twisted ring counter.

Referring again to the schematic the reader will see that we have denoted the flip-flops as FFA, FFB and FFC. The outputs of the flip-flops are labelled as A, B, and C for the Y outputs and A-complement, B-complement, etc. for the Y-complement outputs.

Imagine that we have reset all flip-flops to zero (i. e. all Y outputs will be at logical 0, here equivalent to zero potential). Thus, we will have three outputs at a positive potential: A-complement, B-complement and C-complement. Upon the injection of a pulse into the toggle terminals of the flip-flops, they will switch over to give positive outputs at A, B-complement and C-complement. A table can be drawn up to show which output is in the positive state after each successive input pulse. This table is, in fact, compiled from the truth table Fig. 1A

See Fig. 1B

From this table we can see that each set of positive outputs is repeated once for every six input pulses. In fact, we have the makings of a 6 : 1 scaler or frequency divider. It can be seen that, for each set of positive outputs we can select a combination of two values occuring once only for every six input pulses. Each combination can be correlated with numbers from one to six, depending on the order of occurrence.

See Fig. 1C

The reader will see that each number occurs once only for every six pulses. Thus, we have a device which will count sequentially from one to six - continuously or for as long as it is driven by "clock" pulses.

In order to have a visual readout from the counter we must supply some means of driving incandescent lamps, neons, etc. To do this we must have an amplifier which will turn on the appropriate readout device only when the appropriate combination occurs. This is accomplished by means of an "AND" gate and a transistor lamp driver to display each decimal number from 1 to 6.

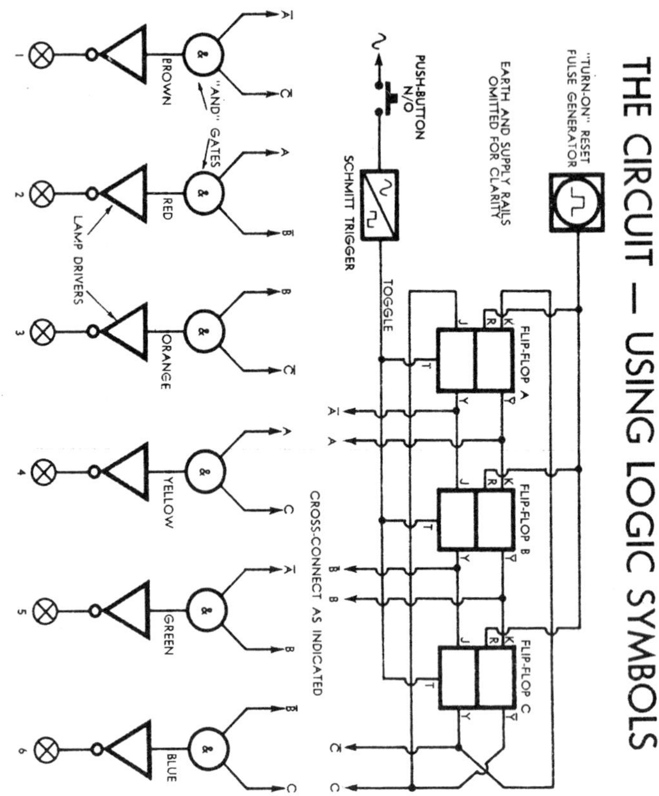

ELECTRONIC DICE

Figure 2: The complete circuit of the dice, without power supply. The three flip-flops are connected as a twisted ring counter. Notice the "twist" in the end-to-end connections. The diagram should be read in conjunction with the explanatory circuits of the individual segments.

An "AND" gate is any circuit or device which performs the function of logical conjunction (product) by delivering an output only when two or more input quantities occur simultaneously. Referring to figure 3, it will be seen that the output from the diode "AND" gate will only be positive when the two inputs are both positive.

Thus, the transistor lamp driver (figure 4) will not be turned on until the corresponding flip-flop combination occurs, allowing the base to rise to a positive potential. The transistor lamp driver is denoted as an inverter amplifier since, when the output from the AND gate becomes positive the transistor saturates, bringing the collector to almost zero potential and allowing the lamp to light.

In order to operate the counter as a "dice" it is fed with pulses of regular frequency. To select a number we need merely interrupt the signal path between generator and counter. Obviously enough, the frequency must be sufficiently high so that the "player" cannot "synchronise" with the flashing lights and prejudice the result. In this unit, the frequency is 50Hz, supplied from the secondary of the power transformer. Thus the lights flicker quite rapidly and there is little chance of anyone succeeding in prejudicing results.

The flip-flops can only be triggered by pulses having very fast rise times or, in the case of these micrologic flip-flops, a fast decay, since they trigger on the negative-going edge of input pulses. Hence the need for an extra micrologic IC. This is a Fairchild Fu1914 "Dual 2-input Gate" device. This is connected so that it functions as a pulse squaring "Schmitt trigger" another bi-stable configuration developed in 1938 by O. H. Schmitt. This provides output square waves with very short rise and decay times.

The ring counter can, of course, be driven at frequencies other than 50Hz and can count reliably at frequencies up to 2MHz. Some interesting effects can be obtained at very low frequencies as the unit slowly counts from 1 to 6.

It may be thought that we have described all the circuitry necessary to make a reliable counter, dice or whatever name you prefer to call this device. There is a catch. The operation of the dece depends on the twisted ring counter cycling through a series of six sets of outputs, as listed earlier. Mathematically minded readers may realise that for three flip-flops there are eight possible combinations of values, two of which do not occur during the normal counting sequence. They are :

$$A \bar{B} C \quad \text{and} \quad \bar{A} B \bar{C}$$

These combinations can occur, however, when the unit is first switched on. And if they do occur, the counter will not cycle in the normal way but will simply alternate between either combination. This can be verified from the truth table for a J-K flip-flop. Thus the counter may not

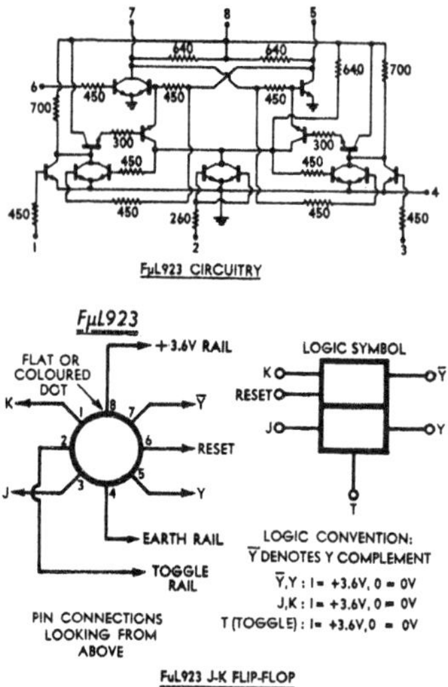

Figure 1: This diagram shows the pin connections of the Fairchild micrologic J-K flip-flop, the internal circuitry and the British Standards symbol for the J-K flip-flop.

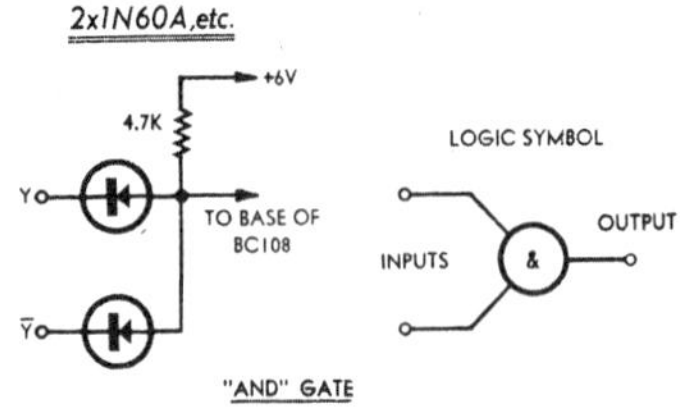

Figure 3: The circuit and symbol of the "AND" gate, six of which are required. Each AND gate responds to one combination of positive outputs from the counter, each combination reoccurring every sixth pulse.

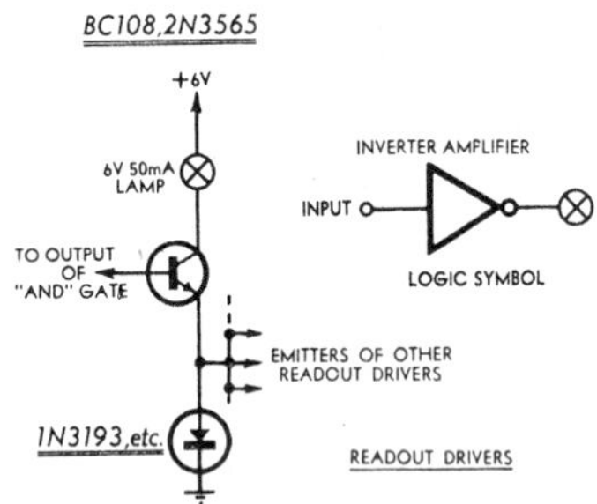

Figure 4: The circuit and symbol of the lamp drivers, six of which are required. The transistor is able to drive the lamp without stress because it is operated in the "switching" mode; high current and high voltage do not occur simultaneously.

simply cycle from 1 to 6 but lamps 1, 3 and 5 or 2, 4 and 6 may flash on alternately. The only way to avoid this is to reset all flip-flops to sero each time the unit is switched on at the mains.

This is accomplished by the "turn-on reset pulse generator" denoted in the schematic as a "clock" pulse generator. This consists of an NPN transistor with its base connected to the positive rail via a resistor and also to ground through a large capacitor. When the power is first connected the capacitor is avirtual short-circuit to ground. This causes the transistor to be initially "cut off", with the collector at supply voltage level. The capacitor then charges up through the resistor, causing the transistor to conduct and bring the collector down to almost zero voltage level. The collector of the transistor is connected to the "reset" terminals of the flip-flops so that the latter are accordingly reset to zero because of the application of the single turn-on pulse. Thus the dice always lights up at "1" when first switched on.

In normal operation the push-button is initially open and no pulses are applied to the counter. Thus only one lamp is alight. Upon pressing the button, the counter cycles and the lamps flicker. To select a number the push-button is allowed to open and one lamp will light. It is, of course, substantially a matter of chance as to which number will result. As with normal dice, it is possible to get repeats of the same number but no past result can prejudice a succeeding try. We carried out a check to see that the unit was not prejudiced towards any particular number, and found that it gave a substantially even distribution of results after a large number of "tosses".

So there it is - a device which is, in itself, rather frivolous but, we think a very interesting application of simple digital circuitry.

CONSTRUCTION :

The components are mounted in a sloping front metal box of a type used in many of our projects and measuring approximately 5" x 5" x 5". The power transformer, which has a 6.3 volt secondary, is mounted on the back of the case. This allows a neater arrangement of components and better accessibility.

The tagboard on the floor of the case at the rear accommodates the power supply components. This supplies 6.3 volts AC via a 39K resistor to the Schmitt trigger, 6 volts DC to the readout circuitry and 3.6 volts DC to the ring counter and associated circuitry. There is also space to mount components for a HT supply for an alternative numerical indicator readout system which we hope to discuss in a forthcoming book. The circuit for the power supply is shown in figure 7.

Although the circuit at present shows only a 6.3V heater transfor as the AC source, readers who may wish later to experiment with a high-voltage readout tube would be well advised to use a small power transformer instead, having available a 150V AC secondary with a nominal 30mA current rating.

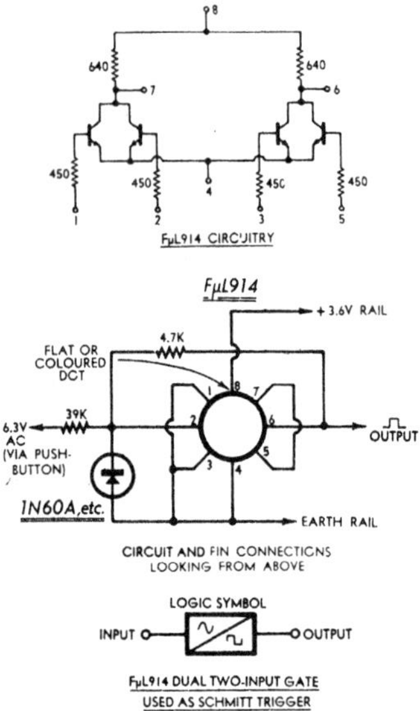

Figure 5: This shows the Fairchild Ful914 Dual 2-input Gate connected as a Schmitt trigger. At top is shown the internal circuitry. The diode is used to protect the Schmitt trigger against "negative-cycles" of the AC input.

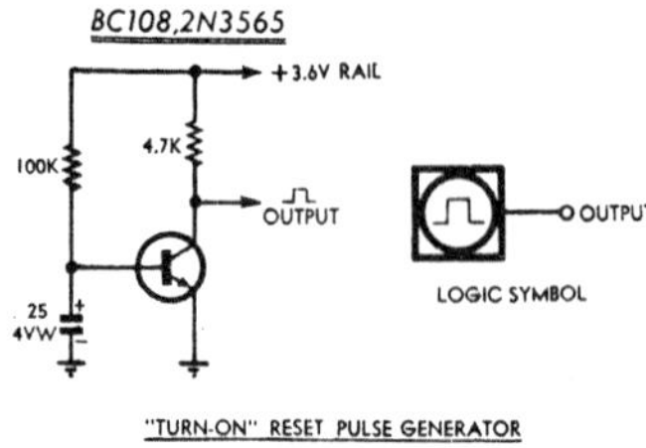

Figure 6: The above circuit configuration generates the reset pulse at "switch-on." For this reason, it is depicted as a "Clock" pulse generator. If this feature was not incorporated in the dice it could "lock-out" at "switch-on."

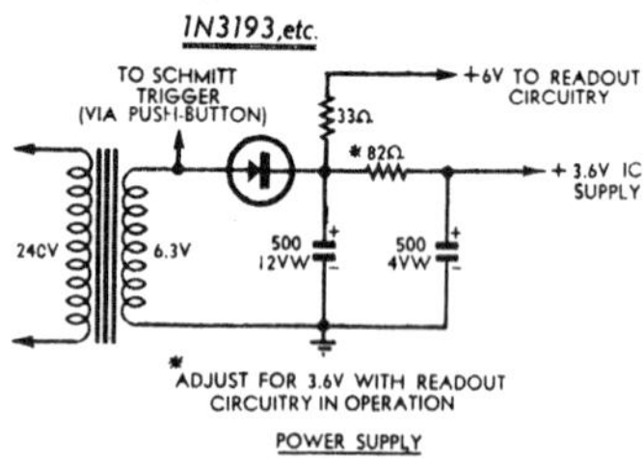

Figure 7: The circuit of the power supply. Readers wishing to construct the version with the numerical indicator tube readout should use a power transformer with an additional 150V AC secondary winding.

For the present, the DC supplies involve the use of a simple half-wave rectifier followed by a "brute force" filter - a single 500uF/12VW electrolytic capacitor. This provides approximately 6 volts to the readout lamps and circuitry. The supply to the ICs is derived via an 82-ohm resistor which must be adjusted to provide 3.6 volts while the readout circuitry is in operation. A further 500uF/4VW electrolytic is used to filter the IC supply, as it was felt the ripple was rather high without it.

The second tagboard on the floor of the case accommodates the three ICs comprising the twisted ring counter, the Schmitt trigger IC, and the "reset" generator transistor which is obscurred on the photograph. The 12 diodes making up the "AND" gates are strung between the appropriate terminals of the flip-flops and a tagstrip. Instead of making connections so that the diodes were physically crossing over each other; with the risk of shorts, each pair of diode connections was made so that leads were as short as possible with the minimum number of crossovers. Then, to avoid confusion, the output of each "AND" gate is colour coded by means of the attached lead according to the resistor colour code, i.e. brown-1, red-2, orange-3 etc.

The readout circuitry is mounted, together with the lamps, on the lid of the case. The tagboard is mounted on a strip of aluminium as shown in the photograph, which, in turn, is attached to the lid by means of the push-button-fixing nuts. This avoids the use of unsightly screws on the front panel.

The lamps we used were 6 volt 50mA PMG No. 2 types and were fitted to the panel by means of close-fitting grommets. There are many 6 volt 50mA lamps suitable, as we have found during research for the current articles on model train lighting.

The row of lamps is positioned about 1½ inches down from the top of the case - so that they do not foul the power transformer. The push-button is mounted about 1½ inches up from the sloping bend in the lid so that it, also, does not foul the circuitry inside the case.

It may be wondered how BC108 or 2N3565 transistors stand up to switching 50mA at 6 volts without exceeding their dissipation ratings. The important point is that the transistors are always in one of two states - either completely "saturated" or "cut-off". When the transistor is saturated, little power is dissipated internally, in spite of the 50mA current, because the voltage across the transistor is less than 1 volt. Similarly, when "cut-off" there is little power dissipated in the transistor.

The emitters of the readout driver transistors connect to earth via a power diode such as a 1N3193 or BA100. This is to ensure that the lamps don't glow slightly when they are supposed to be off. The current of 50mA flowing through the diode induces a voltage of about 0.5 volts across it, thus holding the emitters above earthpotential - and the transistor bases relatively more negative - ensuring that all non-operative transistors are positively "cut-off".

The colour coded leads for the readout circuitry plus the supply and earth return leads and the two leads to the "toss" push-button are laced up in a neat calbe joining the circuitry in the box and on the lid. The readout circuitry and lamp assembly is intended to be completely interchangeable with the numerical indicator readout which we hope to publish in due course.

The push-button is a refrigerator type with "normally open" contacts. And here a word of warning : The first one we tried did not prove to be completely reliable in the prototype, as it did not always make positive contact when pressed. Readers who want better reliability may purchase a micro-switch, but it costs about three times as much.

DECISIONS .... DECISIONS

THIS LITTLE GADGET, CAPABLE OF GIVING RANDOM "YES - NO" ANSWERS CAN BE TREATED AS A HUMOROUS GIMMICK, IF THATS THE WAY IT APPEALS TO YOU. ALTERNATIVELY, IT CAN BE USED TO DEMONSTRATE RANDOMNESS AND PROBABILITY OR TO NOMINATE RANDOM NUMBERS.

"To be or not to be" - that is a question which frequently confronts a busy executive. What is more, the choice between two courses of action is often complicated by the presence of an important client who perhaps would not appreciate the tossing of a coin to decide the issue!

The answer lies in an electronic decision-maker, operated secretly in a drawer, or with due flourish and in full view on the executive table top.

The electronic decision-maker is a simple neon multivibrator, which flashes rapidly from one lamp to the other such that both neons appear to be lit at the same time - indicating also that it is switched on and ready for the big moment. When the decision button is pressed, the device is locked in one state, displaying one light with its corresponding momentous decision - "yes" or "no".

For those more interested in sport than business, a possible use for the electronic decision-maker is to replace the pennies in the traditional game of "two-up". The "spinner" would press the button twice for each "toss". Perhaps a combined unit could be built with two circuits operated by one double-pole push-button switch, although some may argue that a degree of randomness may thereby be sacrified.

The game of two-up has been described by a judge as "a good game, when played fairly". Appropriately, no amount of polishing by the "boxer" can influence the decision of the simple electronic "penny", which can ensure a very fair game!

More seriously, however, the decision-maker can be employed in schools and elsewhere for probability experiments and for generating random numbers in binary form. For these applications, it may be desirable to take extra precautions (to be described later) to ensure equal probabilities for "heads" and "tails".

The probability of a specified result of certain events is the relative frequency of occurrence of that particular result in all events of the same kind. Where appropriate, it is thus the ratio of "favourable" cases to the total number of possible cases.

If all results in a series of experiments are favourable, the apparent probability that all possible results would be so tends towards one, which is the highest probability and amounts to a "certainty". As the number of experiments moves towards infinity, this apparent probability can be assumed to approach the actual probability. On the other hand, if all results in the series are unfavourable, the apparent probability tends to zero and to "impossibility".

Practical probabilities lie between these two extremes, and are expressed as fractions. For example, the probability of a new-born baby being a boy is about 0.51. This is calculated from the records of many births, or "experiments", which show that slightly more boys than girls are born.

But whether or not subject to mathematical definition, the factors which combine to produce any given results are commonly called "chance" - a collective word embracing all the complex causes and circumstances which affect a result.

Some probabilities are easy to calculate, since the possible number of results is fixed. To calculate the probabilities which are relevant to rolling dice, we need to consider an individual die first. Each die has six faces and, if it is symmetrical, each face has an equal chance of being on top when the die comes to rest. The die cannot stand on an edge or on a corner, so there are only six possible results. The probability of throwing a specified number is therefore one in six, or one-sixth.

If a coin is tossed, it is reasonable to expect that the probability of it falling heads is one in two, or 0.5. A coin has little chance of standing on edge, and the possibility of this is as remote that it can safely be discounted.

Some claim, however, that a person with practice can toss a coin so as to give a desired result more often than not. Again, a deformed coin may conceivably tend to land on one side more frequently than the other. In such cases, the probabilities of heads and tails must be verified experimentally.

The experimental method of determining the probabilities of heads and tails is simply to toss a penny a number of times and note the results. If the process is repeated a sufficiently large number of times, the

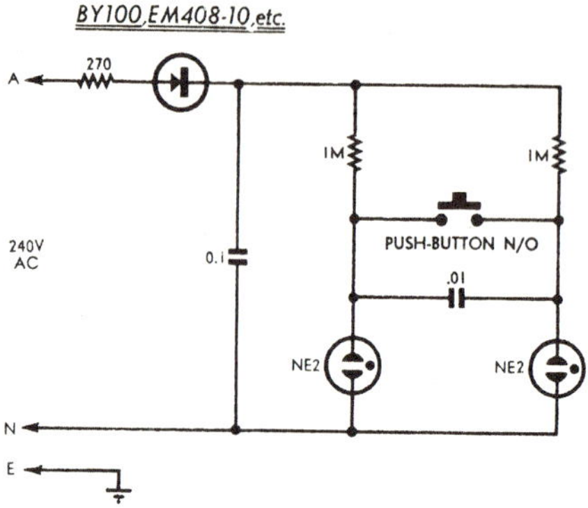

Two ordinary neon oscillators, interconnected as shown, will give adequate symmetry for gimmick use. However, discrete adjustment to balance the yes - no probabilities would be justified, if more serious use is anticipated.

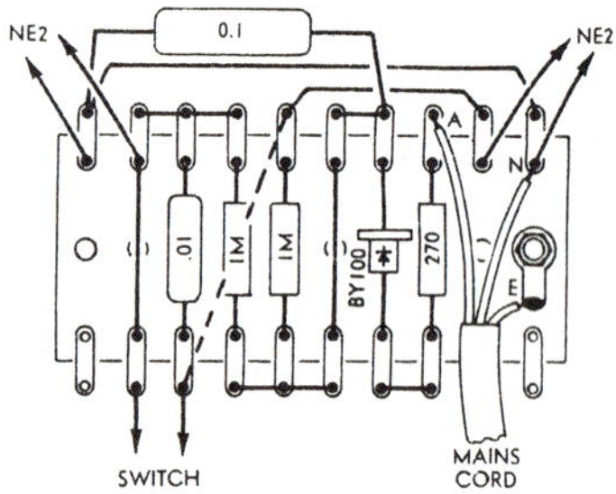

PARTS LIST

1 Eddystone case 4 5/8 x 3 5/8 x 2 1/8 (or smiliar).
2 Lamp bezels.
1 Push-button SPST switch.
2 Neon tubes type NE2 or similar.
1 270 ohms ½-watt resistor.
2 1M ½-watt resistor.
1 .01uF 160V capacitor.
1 0.1uF 400V capacitor.
Mains lead, wires, insulating sleeving, solder. etc.

pattern of results will permit an experimental determination of the probability of heads or tails for the penny (or other device). The larger the number of tosses, or experiments, the more accurate will be the final answer.

The results at first may appear to be contrary to expectation. Out of the first 20 tries, for example, there may be only eight tails. However, by the time there has been 100 attempts, tails could total 48. At 1,000 tails might have reached 498. In other words, with a larger sample the proportion of tails will normally approach half the total to agree with the predicted probability.

If the first four tosses all show heads, for instance, do not immediately doubt all you have read on probability. There will be many continuous runs of heads and of tails in any series of experiments.

The probability of a head turning up is one in two, or 0.5. It does not follow, however, that to maintain the correct ratio the next toss must produce a tail. Many gamblers think that if there has been a run of heads the chance of tails coming up next throw must be increased to give the correct proportion of results. This is not the case. No past result can prejudice a succeeding try.

The probability of 0.5 only means that if we could toss a penny an infinite number of times, the number of heads and tails would be exactly equal. If we take a very large sample, we will obtain approximately equal numbers. With a small sample, however, the ratio of heads to tails can vary considerably.

The probability of a run of say 20 heads can be computed and, although very small, is finite. The probability of a tail being thrown next time is still 0.5 regardless of all the heads.

Consider now a different experiment. If two pennies are tossed (as in two-up), or the decision button of our electronic gadget is pressed twice, what is the probability of two heads appearing?

If the problem is considered intuitively, it would seem reasonable to expect the probability to be one third. There are three possible results : Two heads, two tails, and a head and a tail. One result out of three would be favourable.

But experimental results would almost certainly not agree with this prediction. Out of 1,000 tries, about 250 double headers would turn up instead of the expected 330-odd. Why?

This discrepancy would arise from the assumption that the three possibilities were equal, but this is not so. If the possible results of the double tosses are tabulated, we get the following:

Possibilities	1st Toss	2nd Toss
1	heads	heads
2	heads	tails
3	tails	heads
4	tails	tails

There are in fact four equal probabilities and not three, two of them being indistinguishable when identical pennies are used. The probability of a double head appearing is, therefore, one in four, or 0.25.

Having learned something about chance and probability, we can now take a look at the electronic "decision-maker" and see to what extent its yes-no decisions are truly unbiased.

The circuit is a simple neon multivibrator, with a half-wave rectified mains supply. Any rectifier can be used with a peak inverse voltage rating of 800V or greater. The series resistor should be chosen to limit the initial surge current to less than the peak surge current rating of the rectifier. The reservoir capacitor may be of any value not less than 0.1uF, with a working voltage of 400V or greater.

When the circuit is switched on, the voltage across the reservoir capacitor is applied through the two resistors to the neon lamps. As this voltage increases, one of the neons will strike. When it has struck, the voltage across it falls to its maintaining voltage. There is no initial charge on the capacitor between the two neons, so that the other end of the capacitor must also fall to the maintaining voltage. This is less than the striking voltage of the other neon which is thus temporarily prevented from firing.

In this state, one end of a capacitor is connected to a comparitively low voltage, through the struck neon. The other end is taken to a high voltage through a series resistor, the other neon being effectively an open circuit. The capacitor therefore charges up exponentially towards the high supply voltage.

The increasing charge on the capacitor raises the voltage across the neon which has not struck until it reaches its striking voltage. When this happens, the voltage across the neon falls abruptly to its maintaining voltage. The voltage across the other neon is pulled down by the potential across the capacitor, which is equal to the difference between the striking and maintenance voltage of the second neon. This reduces the voltage across the first neon sufficiently to extinguish it.

The capacitor now charges up in the opposite direction until the first neon fires once more, extinguishing the second neon. The cycle then repeats itself continually. The frequency of this oscillation is high enough for the neon lamps to appear to be lit continuously.

When the "decision" button is pressed, the timing capacitor is shorted out. The two neons are connected together, and the one which has not struck is held at somewhere near its maintaining voltage and is unable to strike. After the button is released, the original cycle starts again.

If the times taken by the capacitor to charge to the striking voltages on the two half-cycles are different the two half-cycles will have unequal time lengths. When the button is pressed, it is more likely to happen during the longer half-cycle, thus increasing the probability of the associated result. The difference is not likely to be worth worrying about if the device is intended as a fun-gimmick but for deliberate observations to do with randomness and probability, more attention to the matter may be warranted.

The voltage range over which the capacitor must charge is equal to the sume of the two differences between striking and maintaining voltages for the neons. This voltage is the same for both half-cycles. However, if the voltages associated with the individual neon lamps are not equal, the capacitor will have to charge up from different minimum voltages towards the fixed supply voltage.

To ensure greater uniformity, therefore, it is desirable to select a pair of neon lamps with approximately equal striking, maintaining and extinguishing voltages. For serious work, it might be advantageous to buy a number of lamps and compare their characteristics experimentally.

One other factor which can make the changing times unequal is a difference in value of the series resistors. It is necessary to select closely matched resistors (perhaps specifying one per cent tolerance) to obtain a high order of randomness of the results.

To perform a rough check of the randomness the mark-space ratio of the circuit should be checked on an oscilloscope, which should preferably be an accurate measuring instrument. A necessary precaution is to isolate the device from the mains with a suitable transformer. If no isolation transformer is available, the device must be correctly polarised with respect to the mains input (as in the circuit diagram) before connection is made between the oscilloscope and the device.

Provided that the two states of the oscillator are carefully balanced, fairly random binary numbers can be generated very simply on the decision-maker by calling one result "O" and the other "1". The button is pressed the necessary number of times, and the results noted. The resultant series of digits may be converted, if required from binary notation to a decimal number.

For example, to obtain random two-figure numbers using a pure binary approach, the button must be pressed seven times for each number. This will give decimal numbers up to 127, so some results producing three-figure numbers will have to be discarded. For example, assuming heads to represent "O" and tails "1", seven successive results from the decision maker are:

1001011

When converted to decimal notation, this gives the decimal number:

75

Alternatively, when generating large decimal numbers, it would be possible to generate each decimal digit separately as a 4-bit BCD group by pressing the decision button four times for each digit. The resultant binary numbers would then be converted into decimal numbers.

For example, assume that in generating a random three-figure decimal number we obtain the following sequence of results:

0111011 1001

If 8421 BCD code is assumed this converts to:

7 3 9

which gives 739 as the required random number. This principle can be extended to generate random numbers of any magnitude.

Our simple circuit was mounted in a solid metal case. However, a plastic case could be used provided that the metal parts of the lamp bezels and switch are earthed. Most of the components are mounted on a pre-wired tag-strip supported on aluminium brackets, one of which also carries the earth connection. The layout can be seen in the accompanying photograph. A wiring diagram of the tag-strip is also given to assist constructors.

If this unit gives pleasure to some and assistance to others, it will have served its purpose.